Environmental Management

Environmental Management

SECOND EDITION

(Fully Revised and Updated)

H.M. SAXENA

RAWAT PUBLICATIONS
Jaipur • New Delhi • Bangalore • Guwahati • Kolkata

ISBN 978-81-316-0340-6

Reprinted, 2016

Published by
Prem Rawat for **Rawat Publications**
Satyam Apts., Sector 3, Jawahar Nagar, Jaipur 302 004 (India)
Phone: 0141 265 1748 / 7006 Fax: 0141 265 1748
E-mail: info@rawatbooks.com
Website: www.rawatbooks.com

New Delhi Office
4858/24, Ansari Road, Daryaganj, New Delhi 110 002
Phone: 011 2326 3290

Also at *Bangalore, Guwahati* and *Kolkata*

Typeset by Rawat Computers, Jaipur
Printed and bound in India

Contents

Preface

Environmental management is a broad and rapidly evolving discipline and now has become a priority throughout the world because of the serious consequences that have either been experienced by mankind or are likely to happen in future. A concern for protection of environment has emerged as a global issue specially after the Stockholm Conference organised by UNEP. A major change come in approach to environmental management is that apart from technological solutions it requires social engineering and creation of new institutions of facilitate social participation for proper resolution of the manifest problems. On the other hand, sustainable development is a key component of environmental management, which has now become a matter of great concern for the improvement of the quality of human life.

This book explores the subject's core themes in 17 well-organised chapters. The first two chapters are related with introduction to environment, environmental management, sustainable development and ecosystems; while in chapter 3 human impact on environment has been discussed. Next three chapters (4, 5 and 6) are related with various facets of environmental management such as approaches and measures, EIA, eco-auditing, standardisation, etc.

In chapters 7, 8 and 9, a detailed analysis of air, water, noise, land and radioactive pollution and their management have been given. The concept of sustainable development and various aspects of sustainability have been discussed in 10th and 11th chapter. Conservation of natural resource and population and eco-crisis are the themes of chapter 12th and 13th. Chapter 14 deals with environmental hazards and their management. The last three chapters are related with global environmental issues, environmental problems and planning in India and environmental ethics and legislation in India.

Many scholars and friends have had a role, either direct or indirect in writing of this book. Acknowledgements have been given throughout the book as a mark of gratitude towards the scholars whose works I have utilised to clarify much of wanted to say.

My wife, Nirmal has been my greatest source of strength. Without her active cooperation and support, this book would have been much harder to write. I am also thankful to all the members of my family and friends for their moral support.

H.M. Saxena

1

Introduction: Environment, Ecology, Management and Development

Environment is not only a determining factor of life on earth but also controller of the existence, growth and development of mankind and all its activities. Primitive man ate wild fruits—hunted and fished—and relied heavily upon nature for his very existence. Even these activities damaged nature to some extent. At that time, man did not concern himself with conservation methods, or in other words, there was no need for it because nature was able to cope with whatever damage was inflicted, quickly restoring its resources. We cannot expect favours from nature; we must take them. Four quite a long time, no thought was given to the possible consequences of such 'taking' from nature. But, the consequences were significant and pervasive. Mankind has long trusted in nature's potential and restorative powers, though there was no reason that these powers were inexhaustible. Only recently man has come to realise the necessity for conservation. Undoubtedly, this is the result of revolution in science and technology which has dramatically increased man's ability to use natural environment and its resources. Nature has increasingly damaged, restorative capability has progressively weakened, and human environment is deteriorating day by day to the point of affecting quality of life.

On the other hand, ecology is the study of interrelationship between all living organisms and environment. Ecology was earlier considered as the domain of biologists. But, today, there can be no such domain; ecology is understood in a much wider context. The term may be applied to nature, man, society and various forms of human activity. Terms such as "human ecology", "social ecology", "ecological technology" and "ecological equilibrium in the environment" have sprung into existence. All these essentially connote the conservation of nature. It implies the need to preserve the existing environment, which is the result both of natural processes and of man's conservation efforts.

The scientific and technological revolution has increased human's impact on the environment. This impact on the environment constantly grows primarily as a result of versatile and all-penetrating industrialisation. The most pronounced forms of such impact are:

- withdrawal of increasingly large amounts of natural resources;
- growing rates of production and consumption of powers;
- profound physical and chemical transformations affecting a vast number of natural substances with the parallel process of discharge into the environment of various wastes, including toxic substances; and
- changes in agricultural and forest areas.

As a result of all these processes, and technological development which increasingly utilises natural resources, significant changes occur in most important properties of the biosphere. Climatic characteristics of the intensively developed areas undergo changes which profoundly affect the natural biogeochemical cycle of materials, upsetting the complex world-wide balance of oxygen and carbon dioxide in the atmosphere. Considerable changes occur in the natural properties of soilsm also. The list of irrevocable changes in the composition of flora and fauna increasingly grows. In brief, the multifarious consequences of modern society's impact on the environment constitute the elements of one of the most pressing problems of today, namely, the problem of protection, conservation and transformation of the environment for the happy life and development of present and future generations.

Everyone of us has a right to live in an environment of quality. Therefore, it is essential not only to understand the environment and complex framework of ecology but also apply our knowledge in sound management programmes to conserve the natural resources and culture. The conservation of resources and maintenance of the quality of environment can be possible only through proper management. Before we discuss these aspects of environmental management in detail, it is necessary to know the nature of environment, ecology and sustainable development.

Environment

Etymologically the term 'environment' means surroundings. Literally, it is an English expression formed by two words, i.e., 'environ' and 'ment' which means 'encircle' or 'all round'. Thus, environment is a complex of many variables which surrounds man as well as all living organisms. Any external force, substance or condition, which surrounds and affects the life of the organism in any way, becomes a factor of its environment. These factors have been variously called as environmental factors, ecological factors or simply factors, and may be living, biotic, as well as non-living, abiotic. The sum of all these living and non-living factors makes the environment of an organism. The place where an organism lives is called 'habitat'. It is also known as 'milieu' which means "total set of surroundings". In fact, habitat is a part of the earth's surface where life originates and develops according to the geographical conditions. Moore, in his book *A Dictionary of Geography* (1967), has defined habitat as "the natural environment of a plant or animal. It is often expressed as one of the main natural regions or one of their sub-divisions, which are recognised in geography as equatorial forests, the tundra". The encyclopaedic definition of environment is as under:

> the complex of physical, chemical and biotic factors that act upon an organism or an ecological community and ultimately determine its form and survival. (*Encyclopaedia Britannica*)

Environmental Factors

As described above, environment is a complex phenomenon of several factors which are the product of the forces and processes of nature, and, are universal and inherent all over the earth. These factors are classified into two groups as under:

(i) *Direct factors*: These are light, temperature, soil, air, water, soil nutrients, etc.

(ii) *Indirect factors*: These are soil structure, soil organisms, altitude, wind, slope, etc.

According to Oosting (1948), the environment is a complex of variable factors or causes, which include: (a) substances (soil, water), (b) conditions (temperature, light), (c) forces (wind, gravity), (d) organisms (plants, animals), and (e) time. Another ecologist Daubenmire (1959) classified these factors into seven coordinate headings as (a) soil, (b) water, (c) temperature, (d) light, (e) atmosphere, (f) fire, and (g) biotic factors. Others have grouped these elements as follows:

(i) climatic or aerial factors,
(ii) topographic or physiographic factors,
(iii) edaphic factors, and
(iv) biotic factors.

While studying environment geographers have also considered the above mentioned factors. These factors are termed as physical factors. The natural environment of any part of the earth's surface can be classified into three cateogries comprising 14 elements:

(a) Abstract elements

(i) Location
(ii) Situation
(iii) Geometrical position
(iv) Size or areal space
(v) Form

(b) Physical elements

(vi) Physiography/landforms
(vii) Climate
(viii) Rocks and minerals
(ix) Soils
(x) Surface waters of the land

(xi) Underground water
(xii) The ocean and its coast

(c) Biotic elements

(xiii) Natural vegetation
(xiv) Animal life

A brief review of these elements is necessary for proper understanding of environment.

Location is an environmental factor, which indicates relative occurrence in space or, in other words, space relationship. Geographical location can be expressed in absolute terms of latitude and longitude relative to the earth's surface as a whole. The latitudinal location also is an indicator of the broad climatic zone in which a specified area/place is located. Whether the place is located in tropical, temperate or cold region, can easily be ascertained simply by its latitudinal location. Apart from this, location in terms of land and sea also affects the local environment. This is known as situation and it means the occurrence of a place with regard to the fixed features of nature. The basic types of situation are: (i) *continental situation* : (a) intermontane situation, (b) midland situation, (c) piedmont situation, and (d) river plain situation; and (ii) *maritime situation* : (a) littoral situation, (b) peninsular situation, (c) insular situation, (d) pelagic situation, and (e) isthmian situation. All these situations have had their impact on the environment in one way or other. Similarly, *size*, i.e., areal space, is also an environmental factor which controls the climate as well as regional environment.

Relief or landform is a main physical factor, which affects the environment variously. The impact of relief can be seen on climate, natural vegetation, soils as well as on settlement of man and economic development. Because of favourable relief, the river valley civilisations have developed in China, India, Mesopotamia, and even today it is a controlling environmental factor for all types of economic development. The various forms of landforms developed in areas like hilly, plain, plateau, arid, glacial, coastal, limestone, etc., regions have had specific natural environment and ecological settings. Mountains are least hospitable to human cultures. The topographical effect on environment can be seen on precipitation, development of snow fields and glaciers, rate of erosion, variations in vegetation types, soils, etc.

Climate of a region is determined by such methodological factors as temperature, pressure, winds, relative humidity, rainfall, evaporation rate, etc. These factors are concerned mainly with the aerial environment of an organism. Climate is not only directed but determines the entire ecological balance of a region. The other environmental factors like natural vegetation, soil, water bodies as well as man's activities are in one way or the other related with climate. The various forms of ecosystem, such as deserts, coasts, glaciers, are the products of climate. The climate is highly affected by man's activities in the form of pollution, which has now become a global problem. The direct effects of temperature and rainfall can be observed on vegetation and animals.

Soil is also an important environmental factor. Plants depend for their nutrients, water supply and anchorage upon the soil. Soil system is indeed very complex and dynamic, undergoing continuous change, and the rate of such change is influenced by a number of other factors of the environment. Soil formation depends upon parent material, climate, vegetation, relief and time. A properly managed soil is an asset; if it is misused, it can be permanently damaged and will create environmental degradation.

Natural vegetation is an index of other environmental factors such as climate, soil, landforms, while by itself it is a factor of environment. It controls the temperature and helps rainfall by increasing humidity. It is also a controller of soil erosion. Forests are assets to mankind and a natural resource of national importance. Great damage has been done to forests after the Industrial Revolution and now the world is facing several problems due to deforestation which has become the main cause of environmental degradation both in developed as well as developing countries.

Among other environmental factors, *water bodies* and *animals* are also important. All these factors of the environment are not independent but are linked with each other in such a way that none can be separated from the other. Therefore, the environment should be taken into consideration in its entirety. The total impact of environment is reflected in all man's developmental activities—economic, social, political or cultural.

Ecology

Ecology is the science of the relations of all organisms to their environment. The two components of nature—organisms and their environment—are not only complex and dynamic but also interdependent. Ecology is a relatively new science, dealing with the various principles which govern such relationships between organisms and their environment.

In 1859, St. Hilaire Isodore Geoffory, a French zoologist, proposed the term 'ethology' for "the study of the relations of organisms within the family and society in the aggregate and in the community". Similarly, a British naturalist, St. George Jackson Mivart, used the term 'hexicology', which he defined in 1894 as "devoted to the study of the relations between the organisms and their environment as regards the nature of the locality...". The term 'ecology' was coined by combining two Greek words, viz., 'oikos' (meaning 'house' or 'dwelling place') and 'logos' (meaning 'the study of') to denote the relationships between the organisms and their environment. Although there is some controversy about the original coining of the term but there is consensus that German biologist, Erust Haeckel first gave substance to this term. He used this term first in 1886 but defined it for the first time in 1870 as follows: "By ecology we mean the body of knowledge concerning the economy of nature—the investigation of the total relations of the animal both to its inorganic and to its organic environment, including above all, its friendly and inimical relations with those animals and plants with which it comes directly and indirectly into contact."

Now, ecology is a systematic and well-developed science. Some of the definitions of ecology are as follows:

> Ecology is defined as the study of the relations of organisms, or groups of organisms to their environment, or the science of the interrelations between living organisms and their environments. (Odum)

Ecology is the science of all the relations of all organisms to their environments. (Taylor)

Ecology is the science which treats the organisms in relation to their environment. (Moore)

Ecology is the science dealing with the relationship of organisms to one another and to other factors that comprise their environment. (*Encyclopaedia Britannica*)

Ecology is the study of plants and animals in relation to their environment. (Haggett)

Ecology is the science of the mutual relationship of organisms to their environment. (Monkhouse and Small)

It becomes clear from the above definitions that ecology is the study of the interrelationship between all living organisms and environment. All its components, i.e., energy, air, water, soil, vegetation, animals as well as man, have mutual relations, responsible for the development of a particular ecological system. The interrelationship of ecological elements has been depicted below in Figure 1.1:

Figure 1.1
Ecological Relationship

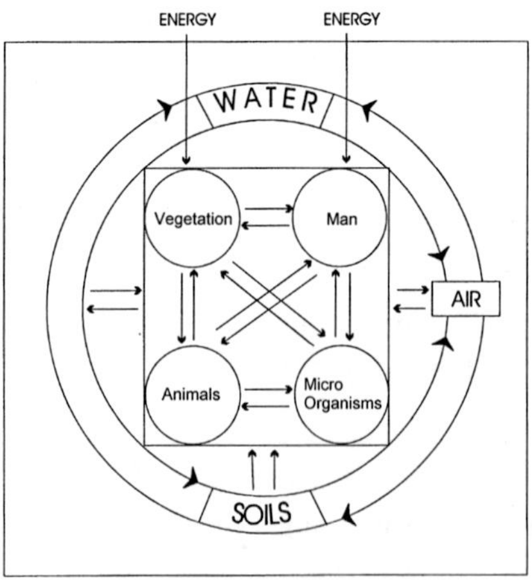

All the elements—biotic and abiotic—are not only interrelated but also function in an orderly manner as a definite system. Thus, *structure and function* provide a base for study of ecosystem. The spectrum of modern ecology can also be defined in terms of levels of organisation. It begins with the groups of organisms, up to ecosystem level. In ecology, the term 'population' is used to include groups of individuals of any one kind of organism. Likewise, community or biotic community in the ecological sense includes all the populations occupying a given area. The community and the non-living environment function together as an ecological system or ecosystem. The present status occupied by ecology is quite evident by its role in the solution of socio-economic and environmental problems such as pollution. Its role in managing water, soil, forest and land resources as well as in agricultural development and conservation of resources has become important. The problem of environmental pollution also needs an ecological approach. In fact, sustainable development is nothing but development within the framework of an ecosystem.

Basic Concepts of Ecology

The basic concepts of ecology are as follows:

(i) All living organisms and their environment are mutually reactive, affecting each other in various ways.

(ii) The environment, which is a complex phenomenon of several interrelated factors, is highly dynamic (i.e., varying with time and space).

(iii) The species puts forth every effort to maintain its uniformity in structure, function, reproduction, growth and development by preservation of its genetic pool.

(iv) Not only does the environment influence the life of organisms, but organisms too modify their environment as a result of their growth, disposal, reproduction, death, decay, etc.

(v) The concept of 'biome' was put forward by Clements and Shelford, who stated that all plants and animals are related to each other by their coaction and reaction on the environment. According to them, under similar climatic conditions, they may simultaneously develop more than one community, some reaching the climax stage, others under different stages of succession. This complex of several communities in any area, represented by an assemblage of different kinds of plants, animals, etc., sharing a common climate, is called biome.

(vi) Each population occupies a specific niche, a unique functional position with respect to other organisms with which it interacts.

(vii) There exist varying degrees of neutral interactions among organisms, both at inter-specific and intra-specific levels.

(viii) The chemical components of the ecosystem move in defined cycles known as 'biochemical cycles'.

(ix) The successful growth of an organism is governed by limiting factors. The minimal and maximal levels of tolerance for all ecological factors of a species vary seasonally, geographically and according to the age of the population.

(x) The exploitation of ecosphere by man's activities is the main cause of its widespread disruption. It leads to a setback of the state of development and reduction in the stability of the ecosystem.

Management of Environment and Sustainable Development

Environmental management is the process of allocating natural and man-made resources so as to make optimum use of the environment in satisfying not only the present basic human needs but of the coming generations also. This management implies an element of conscious choice from a variety of alternative proposals and furthermore that such a choice involves purposeful commitment to recognised and desired objectives. Environmental management is not merely a management of environment but it is essentially the management of various activities with intolerable constraints imposed by the environment itself and with full consideration of ecological factors. Thus, it involves environmental planning, conservation of resources, environmental status evaluation, and environmental legislation and administration. The focus of environmental management is on implementation, monitoring and auditing; on practice and coping with real-world issues, rather than theoretical planning. A close integration with environmental planning is desirable. Thus, as stated earlier, environmental management is a field of study dedicated to understanding human-environment interactions and the application of science and commonsense to solving problems. The characteristic features of the environmental management are:

— it deals with a world affected by humans;
— it supports sustainable development;
— it demands a multidisciplinary approach;
— it has to integrate different development viewpoints;
— it concenrns with short-term and long-term planning as well as from local to global scale; and
— it seeks to integrate natural and social science, policy making and planning.

During the last three decades too much awareness has been developed regarding environmental protection and quality of life. The dictionary of environment is renewed regularly with new terminologies like clean technology, environmental auditing, environment-friendly products, environmental impact assessment, environmental resource conservation, etc., added. But all these aspects have been converged when the wider concept of environmental management has been emerged and also accepted as a tool for sustainable development. Environmental management, as defined in Goudie (1994), "provides resources from the bioenvironmental systems of the planet but simultaneously tries to retain sanative, life-supporting ecosystems. It is therefore an attempt to harmonise and balance the various enterprises for his own benefit."

Time has now come when our policy makers as well as society should aim to protect, conserve and regulate the development in such a way that it will not create any adverse effect on ecosystem and needs of the people can also be fulfilled. Throughout the world, particularly in developing countries, these is an urgent need for the management of the total environment. In the first instance environmental management must do three things: (i) identify goals, (ii) establish whether these can be met, and (iii) develop and implement means to do what it deems possible. A simple scheme for environmental management has been depicted in Figure 1.2.

Figure 1.2

A General Scheme for Environmental Management

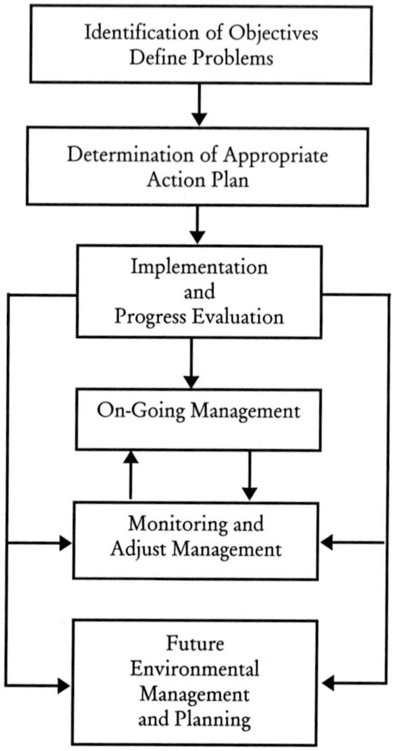

Thus, environmental management is an approach which integrates ecology, policy making, planning and social development. Its main objectives include:

— to prevent and solve environmental problems;
— to establish limits;
— to develop research institutions and monitoring systems;
— to warn threats and identify opportunities;
— to suggest measures for resource conservation;
— to develop a strategy for the improvement of quality of life;
— to suggest long-term and short-term policies for sustainable development; and
— to identify new technology for sustainable development.

In brief, environmental management is necessary for environmental planning which implies the optimal utilisation of the earth's resources and preservation of the quality of environment for the healthy growth of society.

The term 'sustainable development' was first used at the time of the Cocoyoc Declaration on Environment and Development in early 1970. Since then, it has become a matter of great concern not only to the international and national organisations but also to the policy makers. In 1987, the World Commission on Environment and Development has published a report *Our Common Future* emphasising the need for sustainable development and defined it as : "Sustainable development is the development that meets the needs of the present without compromising the ability of future generations to meet their own needs". Others refer to it as "improving the quality of human life while living within the carrying of the supporting ecosystem". Now, sustainable development, also referred as 'eco-development', has been accepted as a concept of development by all the nations of the world. *Agenda 21,* adopted in the UNCED (3–14 June, 1992) at Rio de Janeiro, is a programme of action for sustainable development, also known as 'Rio Declaration' on Environment and Development.

References

Encyclopaedia Britannica, Vol. III, 1980, p. 780.

Goudie, A. (ed.) (1994) *An Encyclopedic Dictionary of Physical Geography,* Blackwell, Oxford, p. 181.

Haggett, P. (1975) *Geography: A Modern Synthesis,* Harper and Row, London, p. 111.

Moore (1967) *A Dictionary of Geography.*

Monkhouse, F.J. and Small, J. (1980) *A Dictionary of Natural Environment,* p. 38.

Odum, E.P. (1971) *Fundamentals of Ecology,* p. 43.

World Commission on Environment and Development (1987) *Our Common Future,* Oxford University Press.

2

The Ecosystems

The whole earth and its parts, viz., land surface, water bodies, atmosphere as well as living organisms, are functioning in an integrated system, which not only directs but also determines the pattern of life. It also indicates relationship of environment with living organisms and also provides a basis for environmental planning and management. Each and every part of the earth's surface is having its own ecosystem but at the same time it is also a part of a larger system, which may cover the whole of the continent, parts of various continents or even whole of the earth. The single giant ecosystem is the 'biosphere'. The term 'ecosystem' was coined by A.G. Tansley in 1935, who defined it as "the system resulting from the integration of all the living and non-living factors of the environment". He further stated that "the whole system includes not only the organism complex but also the whole complex of physical factors forming what we call the environment of the biome—the habitat factors in the widest sense. It is the systems so formed which are basic units of nature on the face of the earth".

Although the concept of ecosystem is much older than the term itself, many terms parallel to it have been used by bio-scientists, such as *biocoenosis* (Mobius, 1877), *microcosm* (Forbes, 1887), *holocoen* (Frienderichs, 1930), *biosystem* (Thienemann, 1939), *geobiocoenosis* (Sukachev, 1944), *ecosphere* (Cole, 1958), etc. However, the term 'ecosystem' is now accepted and most preferred, where 'eco' implies the environment, and 'system' means interdependent, interacting complex.

According to Monkhouse and Small, ecosystem is "an organic community of plants and animals viewed within its physical environment or habitat."[1] The habitat or physical environment controls the whole world of organic community, viz., plants and animals. With the change in physical environment, the types of plants and animals too have changed. Similar views have been expressed by Haggett: "Ecosystems are ecological systems in which plants and animals are linked to their environment through a series of feedback loops."[2]

Strahler and Strahler have defined ecosystem as "the total assemblage of components entering into the interactions of a group of organisms". He further elaborated that "to the geographer, ecosystems are the part of the physical composition of the life layer."[3]

In *Encyclopaedia Britannica* ecosystem has been defined as "a unit that includes all the organisms (biological factors) in a given area interacting with the environment (physical factors) so that a flow of energy leads to a clearly defined trophic (nutrient requiring) structure, biotic diversity, and material cycles (i.e., exchange of materials between living and non-living sectors)."[4]

In fact, our earth has a giant ecosystem in the form of the biosphere, where abiotic and biotic components are constantly acting and reacting upon each other, bringing forth structural and functional changes in it. The biosphere is a vast ecosystem and difficult to study as a whole, therefore, for the convenience of study, it can be divided into the following major sub-divisions:

1. *Terrestrial* as forest, grassland, desert, mountain, etc.
2. *Aquatic* (water-open): (a) *fresh water* which may be *lotic* (running water like spring, stream, river) or *lentic* (standing water like lake, pond, swamp, pools, etc.); and (b) *marine* such as deep sea or shallow sea, delta, etc.

Components of the Ecosystem

The two major components of ecosystem are:

— the *biotic* components comprising all living organisms, and
— the *abiotic* components including the physical (non-living) environment.

Biotic Components

In an ecosystem organisms are usually divided into two categories on the basis of their nutritional (trophic) relationship. These are: (a) autotrophic component, and (b) heterotrophic components.

(a) *Autotrophs* are those organisms which can produce their own food. These are green plants (with chlorophyll) and certain bacteria, which obtain their energy from the sun and through photosynthesis produce food. Since these organisms produce food for all the other organisms, they are known as producers.

(b) *Heterotrophs* are those organisms which depend directly or indirectly upon the autotrophs for their food, thus known as consumers. The consumers are of two types:

   (i) *Macroconsumers*, also known as herbivores, carnivores or omnivores. *Herbivores* are the plant eaters, while carnivores are animal eaters. On the other hand, omnivores eat all kinds of food. They all are phagotrophs which include chiefly animals that ingest other organic and particulate organic matter. These are also known as primary consumers.

   (ii) *Microconsumers*, also known as decomposers or osmotrophs. They are decomposers because of their role in decomposition of dead organic matter. They secrete digestive enzymes to break down food into simpler substances and then absorb the digested food.

Functionally, the biotic components can be divided into three types, viz., producers, consumers and decomposers. Producers produce their own food like plants, which are primary producers, while man and animals obtain food from plants. Similarly, consumers are also primary, secondary or tertiary. Man, cow, goat, etc., are primary consumers, while meat eaters are secondary consumers. Tertiary consumers are those animals which eat meat such as snakes (who eat frogs), peacocks (who eat snakes), etc. Among decomposers those bacteria, fungi, etc., have been included which are responsible for decomposition of a product and release inorganic nutrients in environment.

Abiotic Components

The physical environment with its several interacting variables constitutes the abiotic component of ecosystem. It consists of (i) the lithosphere, i.e., the solid mineral matter of the earth, (ii) the water in the oceans, rivers, lakes, etc., or in other words, the hydrosphere, (iii) the gaseous mixture in the air (the atmosphere), and (iv) the radiant solar energy.

Figure 2.1 is a generalised depiction of the structural elements of an ecosystem and their functional linkages.

Figure 2.1

Ecosystem Model Showing its Structure and Function

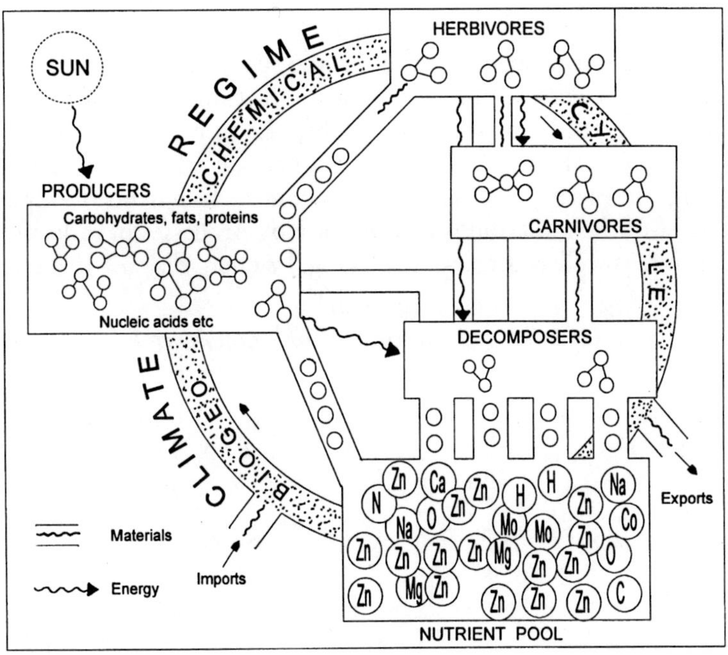

Ecosystem Functioning

The functional pattern of an ecosystem is a very important and basic aspect of study, because all its components are dynamic in one way or the other and are responsible for the creation of an unique state of man-environment relationship as well as a habitat, different from other regions. Ecosystem functioning can be studied under three heads, viz., energy flow, productivity and biogeochemical cycling.

Energy Flow

Energy is defined as the ability to do work and is the essence of life. Without energy transfers, there could be no life and no ecological systems. The energy used for all the life processes is derived from solar radiant energy. The behaviour of energy is described by the laws of thermodynamics: (i) energy may be transformed from one type into another but is never created or destroyed, and (ii) no process involving an energy transformation will spontaneously occur unless there is a degradation of the energy from a concentrated form into a dispersed form, etc.

The radiant energy produced in the sun travels through space in the form of waves. But only a small fraction of solar radiation reaches the earth to provide energy for the biotic components of the ecosystem. Figure 2.2 indicates the fate of radiant energy reaching the earth's atmosphere. It becomes clear from the figure that most of the radiation is lost in space by processes of reflection, absorption as well as through scattering. Its energy is greatly altered

Figure 2.2

Radiant Energy Reaching Earth's Atmosphere

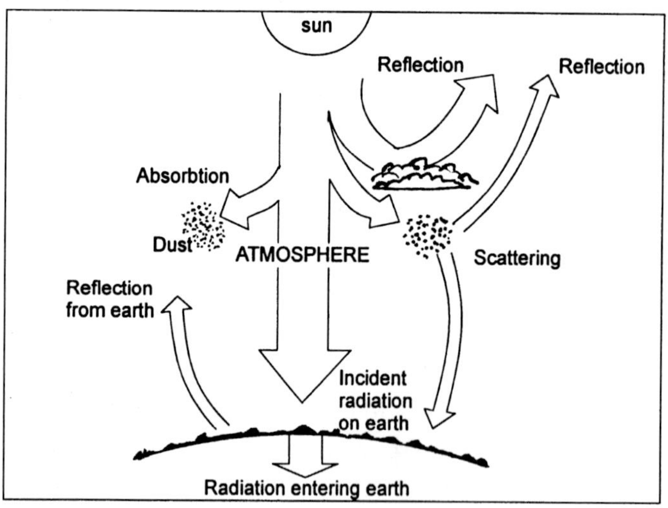

as it passes through cloud cover, water and vegetation. The daily input of sunlight to autotrophic layer of an ecosystem varies mostly between 100 and 800 to 300–400 g cal per cm^2 (= 3000 to 4000 kcal per m^2) in the temperate zone. The total radiation flux within different strata of the ecosystem varies from season to season as well as with nature of the earth's surface, thus controlling the distribution and response of organisms accordingly.

The energy reaching the earth's surface is transformed and/or absorbed by plants and other organisms. It is used by the green plants during photosynthesis by converting the light energy to chemical energy and making it available to other organisms as food. Such energy transfers along the food chain were studied for the first time by Lindeman in 1942. He stated that all the functions of an ecosystem could be explained in terms of energy by the knowledge of two attributes of each trophic level—the level of energy storage and the efficiency of energy transfer. Odum (1957), Golley (1960),

Figure 2.3

Energy Flow in a Food Chain

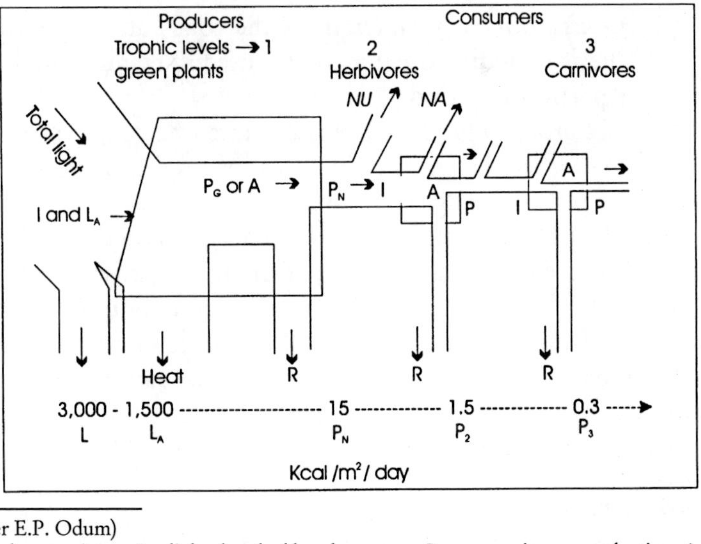

(After E.P. Odum)

(I-total energy input; L_A-light absorbed by plant cover; P_G gross primary production; A-total assimilation; P_N-net primary production; P-secondary (consumer) production; *NU*-energy not used (stored or exported); NA-energy not assimilated by consumers (egested); R-respiration; bottom line in the diagram shows the order of the magnitude of energy losses expected at major transfer points, starting with a solar input of 3,000 Kcal per metre per day.

Slobodkim (1959, 1960, 1962), Teal (1962) and Kozlovsky (1968) have made further detailed investigations. The process of energy flow in an ecosystem has been depicted by Odum in a simplified diagram as shown in Figure 2.3, where the 'boxes' represent the trophic levels and the 'pipes' depict the energy flow in and out of each level. It becomes clear from this simplified model that the energy flow is greatly reduced at each successive trophic level from producers to herbivores and then to carnivores.

Productivity

The productivity of an ecosystem refers to the rate of production, i.e., the amount of organic matter accumulated in any unit of time. It is of three types, viz., primary, secondary and net productivity.

(i) *Primary productivity* is defined as "the rate at which radiant energy is stored by photosynthetic and chemosynthetic activity of producer organisms (chiefly green plants) in the form of organic substances which can be used as food materials."[5] Primary productivity is of two types:
   (a) *Gross primary productivity* is the total rate of photosynthesis including organic matter used in respiration during a particular period.
   (b) *Net primary productivity* is the rate of storage of organic matter in plant tissues in excess of respiratory utilisation by plants during the measurement period.

(ii) *Secondary productivity* is the rate of energy storage at consumer level. It actually remains mobile from one organism to another and does not live *in situ* like primary productivity.

(iii) *Net productivity* refers to the rate of storage of organic matter not used by the consumers. Thus, it is the rate of increase of biomass of the primary producers, which has been left out by the consumers.

Biogeochemical Cycling

"The chemical elements, including all the essential elements of protoplasm, tend to circulate in the biosphere in circular paths from environment to organism and back to the environment. These move, in a more or less circular path, are thus known as *biogeochemical cycles*".[6] The growth of life process requires about 40 elements. Hydrogen, carbon and oxygen are the basic elements, while nitrogen, phosphorus, potassium, calcium, sulfur, magnesium and iron are the other important elements. All the elements continue to cycle at the ecosystem level; they are also an integral part of the larger and/or global cycle. The cycling of various elements can be grouped into two categories, viz., (i) gaseous cycles of elements like hydrogen, carbon, oxygen, nitrogen, sulfur; and (ii) sedimentary cycles of those elements which are present in the soil and organisms. The process of biogeochemical cycles can be understood through following examples:

Hydrological Cycle

The hydrological cycle or water cycle is the most important one. It involves interchange of water between earth's surface and the atmosphere through rainfall and evapotranspiration. The water from water bodies like oceans, lakes, rivers, etc., gets evaporated by solar energy. These water vapours, after cooling and condensation form clouds, and result in rainfall, snowfall, etc. A large part of the

Figure 2.4
Hydrological Cycle

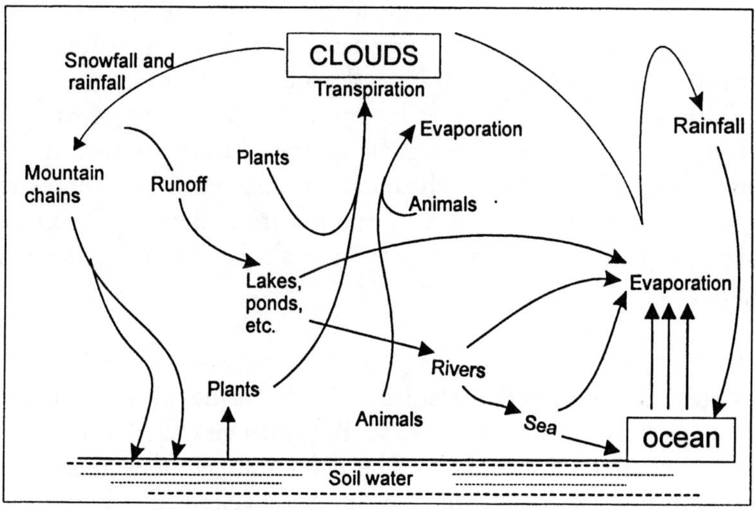

rainfall occurs over oceans, and a sizeable part of the water vapours goes back to the oceans via rivers and streams. Some part of it infiltrates into the soil to become soil or underground water. A small quantity of water is also absorbed by the plants and consumed by other animals. The same is released during respiration and transpiration of plants. Thus, there is a continuous cycling of water as shown in Figure 2.4.

Carbon Cycle

Like the hydrological cycle, the carbon cycle is also very important for man. The small amount of carbon dioxide (CO_2—0.03%) in the atmosphere is the only source of all carbon that passes through the organisms along the food chains. As shown in Figure 2.5, carbon moves from the atmospheric pool to green plants (producers), then to animals (consumers) and finally to bacteria and other micro-organisms (decomposers) that return into the atmosphere, through decomposition of dead organic matter. The carbon dioxide has the unique property of absorbing infra-red radiation and its small quantity helps in keeping the earth warm. A part of this cycle also operates in the ocean.

Oxygen Cycle

The oxygen constitutes 21 per cent of the atmosphere. It also occurs in a bound state as oxides and carbonates in rocks and in water. The plants release oxygen in photolysis of water during photosynthesis. Again, gaseous oxygen is used in respiration of all the organisms and in the oxidation of the organic matter. Another phase of the oxygen is the ozone layer of the outer stratosphere which serves an important function of protecting the life from the ionising ultraviolet waves. The pattern of oxygen cycle has been depicted in Figure 2.6.

Nitrogen Cycle

Nitrogen is an essential element of all forms of life and in atmosphere, its percentage is 79. But nitrogen is never taken directly from the atmosphere. The chief sources of nitrogen for plants are nitrates in the soil. During the cycle, the reduced and

Figure 2.5
Carbon Cycle

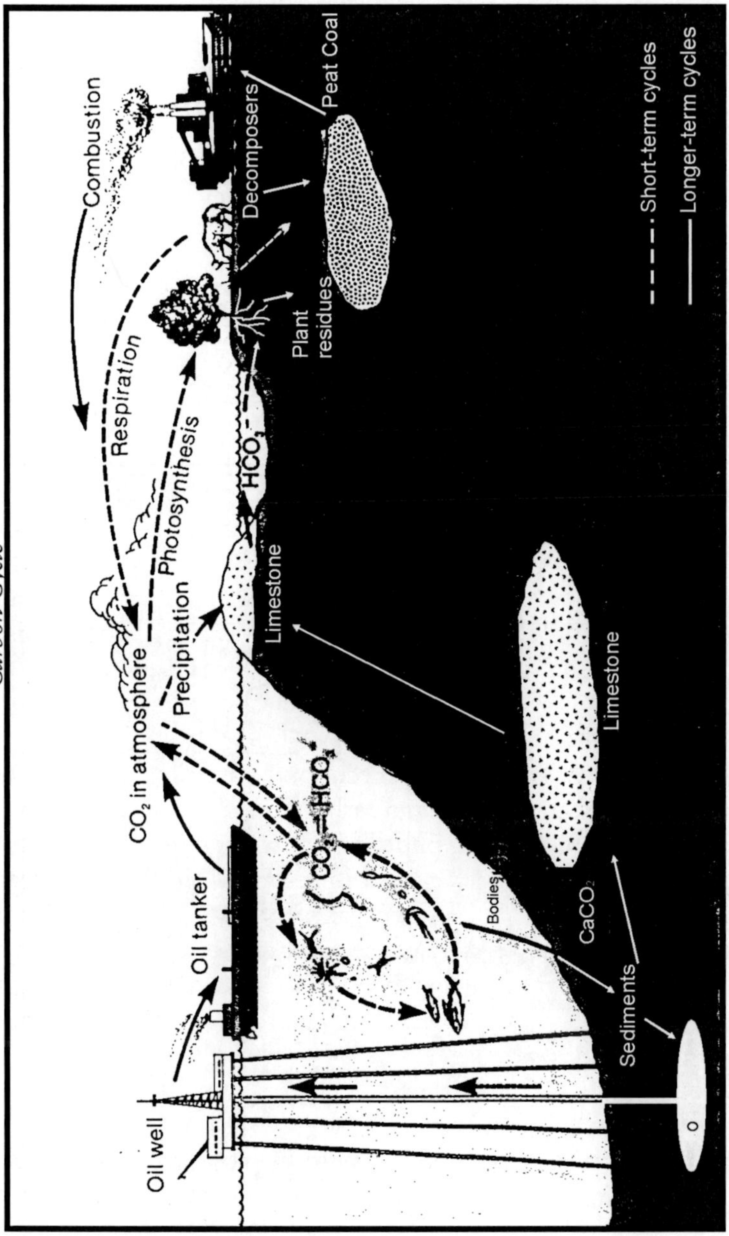

Figure 2.6
Oxygen Cycle

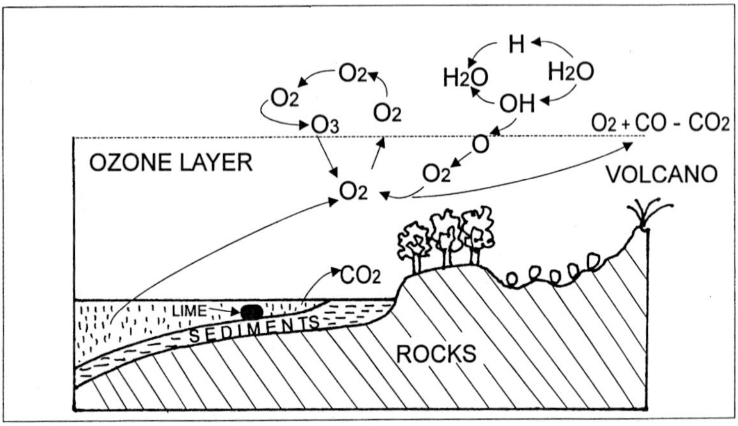

O_2 = Oxygen, O_3 = Ozone, H_2O = Water, CO = Carbon Monoxide, CO_2 = Carbon Dioxide

Figure 2.7
Nitrogen Cycle

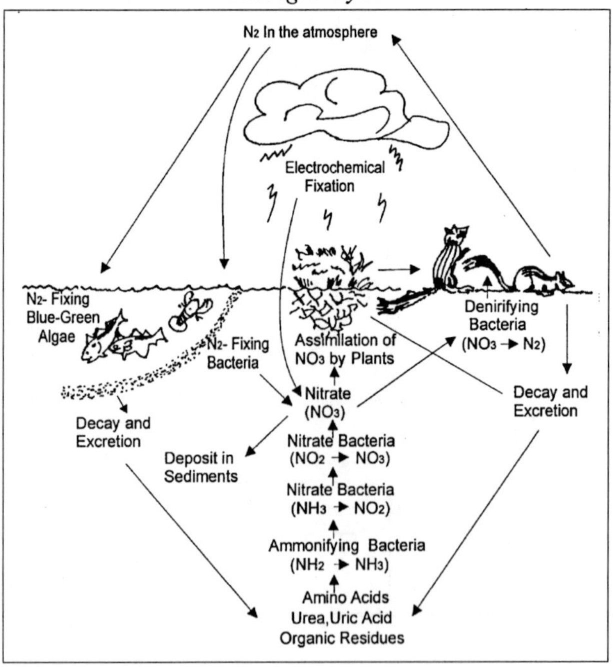

oxidised forms are involved at one or other stage. The chief symbiotic nitrogen fixers are bacteria belonging to the genus *Rhizobium* found in the root nodules of legumes, while asymbiotic nitrogen fixers are some blue-green algae. Plants in turn are eaten by animals. The dead organic matter formed due to death of plants and animals is decomposed by bacteria, fungi, etc., which releases nitrogen either in free state to atmosphere or as ammonia gas. Ammonia gas may also reach the soil as nitrates through the activity of nitrifying microbes. The whole pattern of the nitrogen cycle becomes clear from Figure 2.7.

Phosphorus Cycle

This is a sedimentary cycle and a simpler one, in comparison to nitrogen cycle. As shown in Figure 2.8, phosphorus, an important and necessary constituent of protoplasm, tends to 'circulate', the organic compounds being broken down eventually to phosphates which are again available to plants. The great reservoir of

Figure 2.8
Phosphorus Cycle

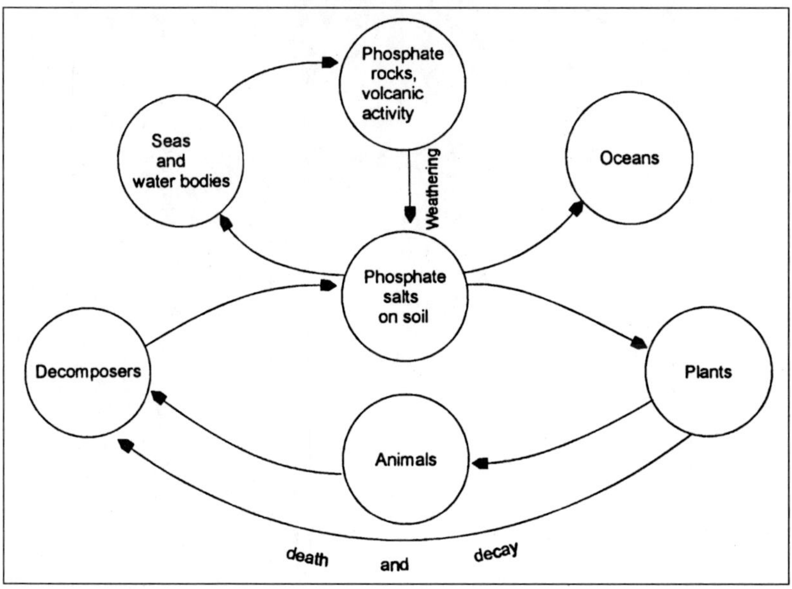

Figure 2.9
Sulfur Cycle

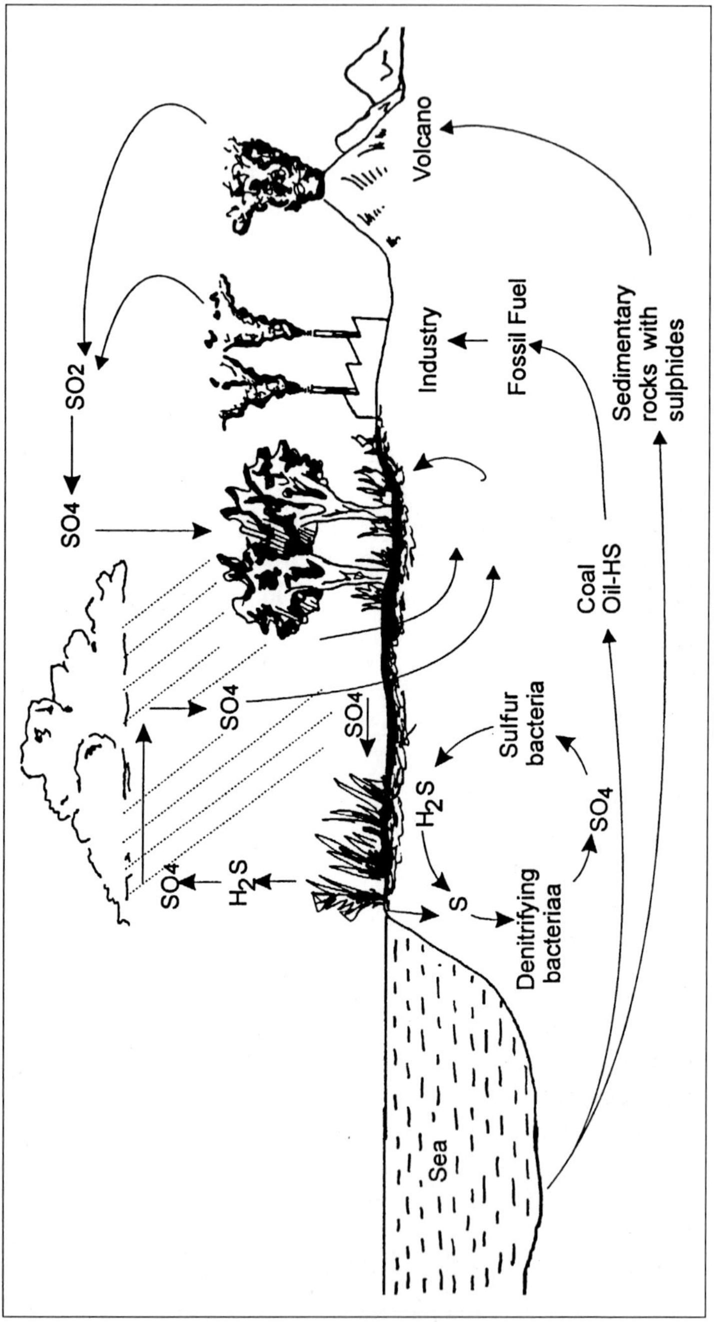

phosphorus are rocks or other deposits which have been formed in past geological ages. These erode and release phosphates to ecosystems. A major portion of phosphates goes into the sea in the form of sedimentation. Sea birds also have a role in returning phosphorus to the cycle from sea to land. The general pattern of phosphorus cycle is more or less similar to that of other minerals.

Sulfur Cycle

The sulfur cycle, as shown in Figure 2.9, links air, water and soil, where microbes play an important role. The reserve pool of sulfur is the soil. It is made available to the plants in the soil by the activity of sulfur bacteria. Some quantities are added to the atmosphere by the burning of the fossil fuels. Later, sulfur dioxide and hydrogen sulfide return to the soil as sulfates or sulfuric acid with the rain. In tissues of organisms, sulfur is in the form of proteins. The decomposition of proteins releases sulfur. The sulfur cycle is a good example of the interaction and complex biological and chemical regulation between different mineral cycles.

Apart from above mentioned cycling of elements, other elements too follow an almost similar pattern of cycling. The cycles of these elements are usually restricted to the same ecosystem but some of them are widespread and a part of the global system.

Major Ecosystems

In the world, there are several ecosystems working at macro or micro level. As pointed out earlier, the biosphere is the biggest ecosystem which combines all the ecosystems of the world. But with the difference in physiography, climate, natural vegetation, soil and water bodies, separate ecosystems have been developed as depicted in Figure 2.10.

The above classification of ecosystems is more geographical, but from the approach of 'habitat' there are several micro-ecosystems working in a geographical region. The regions represent the biomes, which has been defined by Odum thus: "Regional climates interact with regional biotic and substrate to produce large, easily recognisable community units, called biomes."[7] These are primarily the biotic zones also referred to as

Figure 2.10
Basic Types of Ecosystems

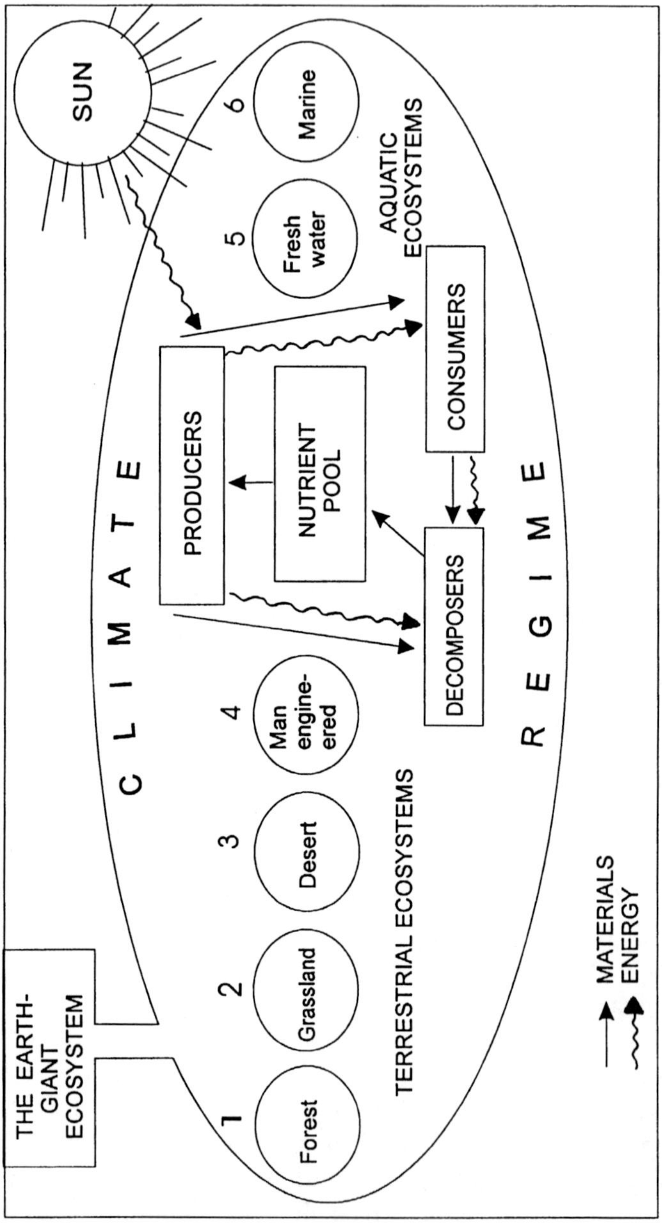

major life zones. Table 2.1 indicates the characteristics of the major environmental zones/biomes of the world whereas the major areas have been depicted in Figures 2.11, 2.12 and 2.13.

The major ecosystems are described as under:

Fresh Water Ecosystem

Fresh water habitats can be divided into two categories:

(i) Standing water or lentic (calm)—lake, pond, swamp or bog.
(ii) Running water or lotic (washed)—river, spring, stream.

Although fresh water habitats occupy a small portion of the earth's surface, they are of great importance to humankind because they provide drinking water as well as water for domestic and industrial needs. Odum has defined fresh water ecology as: "Fresh water ecology emphasises the organisms in environment relationship in the freshwater habitat in the context of the ecosystem principle."[8]

A pond is a good example of fresh water ecosystem, which exhibits a self-sufficient, self-regulating system. A pond is a place where living organisms not only live but also interact with abiotic and biotic components, thus forming an ecosystem which is different from other systems. Similarly, lakes, swampy regions and delta regions of rivers have their own ecosystems in which producers, consumers and decomposers interact and are responsible for the unique ecosystem of each.

Marine (Ocean) Ecosystem

The marine ecosystem is different from fresh water ecosystem mainly because of its salty water and also because (i) the sea covers 70 per cent of earth's area, (ii) the sea is deep, (iii) the sea is continuous, and (iv) the sea water is in continuous circulation. According to Odum, "marine ecology emphasises the totality or pattern of relationship between organisms and the sea environment."[9] In a marine ecosystem, the ecology of shallow and deep waters as well as estuarine part are different from each other. Each ocean in itself represents a very large and stable ecosystem.

Table 2.1

Major Environmental Zones/Biomes of the World

Type	*Zone Name*	*Percentage of Land Surfaces*	*Main Areas*	*Dominant Vegetation Cover*	*Main Man-made Changes in Vegetation*	*Main Precipitation Characteristics*	*Main Thermal Characteristics*
Forested	Equatorial	8	Amazon basin in S. America , Indonesia, S. Asian Peninsulas, Congo basin in Africa	Natural broad-leaved evergreen forest; wide variety of species; swamp forests on floodplains and coast	Very variable: ranges from extensive clearing and cultivation (e.g., in Java) to little impact (e.g., in Amazonia); low to very high population densities	High rainfall (over 100 cm or 39 in) throughout year; heaviest at equinoxes	Uniformly high temperatures; little seasonal variations
	Midlatitude (Peripheral)	7	Europe, E. China, E. United States	Broad-leaved, deciduous, and mixed forests, merging with warm, temperate, evergreen forests on eastern periphery	Very extensive clearing and cultivation; medium-to-high population densities throughout	Moderate precipitation (75-100 cm or 30-39 in) all year; heaviest in winter or autumn on western periphery, in summer maximum in eastern warm temperate area	Cool to warm temperate; seasonal range increases with continentality

Contd...

Contd...

	Boreal	14	Russia and Scandinavia, Canada, Alaska, N.W. United States	Needleleaf forests; relatively uniform stands with small number of species (e.g., spruce, fir, pine, larch)	Limited clearing on equtorward fringes; very low population densities	Light precipitation (25-50 cm or 10-20 in) mainly in summer	Short, cool summers; very large annual temperature range
Intermediate	Savannah	24	African tropics, S. America, S.E. Asia	Ranges from open, tall-grass savannah to deciduous monsoon forest; gallery forest along stream systems	Burning, grazing, variable clearing, and cultivation; high population densities on flood plains in monsoon Asia, otherwise low	Variable rainfall (25-200 cm or 10-79 in) with pronouced spring or summer maximum	Warm; small seasonal variations
	Mediterranean	1	Lands around Mediterranean Sea, S. Australia	Evergreen drought-resistant hardwoods and shrubs	Extensive clearing and cultivation, especially in lands around Mediterranean Sea; variable population density	Low to moderate rainfall (50-75 cm or 20-30 in) with pronounced summer drought	Warm temperate; moderate annual range

Contd...

Contd...

	Midlatitude grasslands	9	C. Asia and E. Europe, C. North America, E. Australia	Grasslands varying from tall-grass prairie to short-grass steppe with decreasing humidity	Hunting and grazing; main settlement and cultivation in last 150 years; low population density	Low to moderate rainfall (30-60 cm or 12-24 in), mainly in spring and summer; considerable year-to-year variation	Very strong seasonal variations; cold winters dominated by invasions of polar air
Barren	Arid and semi-arid	21	C. Asia, Sahara and S.W. Asia, C. and W. Australia	Widely dispersed, drought-resistant shrubs; salt flats, plantless sand, and rock deserts	Little impact outside small irrigated areas	Very low rainfall (0-25 cm or 10 in) with considerable year-to-year variation	Very high summer temperatures; seasonal variations range from moderate in tropics to very high in midlatitudes
	Tundra	5	N. Canada and Alaska, Russia and N. Scandinavia	Low herbaceous plants, mosses, and lichens	Little impact	Low annual totals (10-40 cm or 4-16 in) with late summer or autumn maximum; light winter snowfall	Severe cold; short, cool summers
	Polar	11	Antarctica, Arctic	Ice caps; no plant life	No impact	Low annual totals; little detailed precipitation data	Extreme cold; no months with above-freezing average temperatures

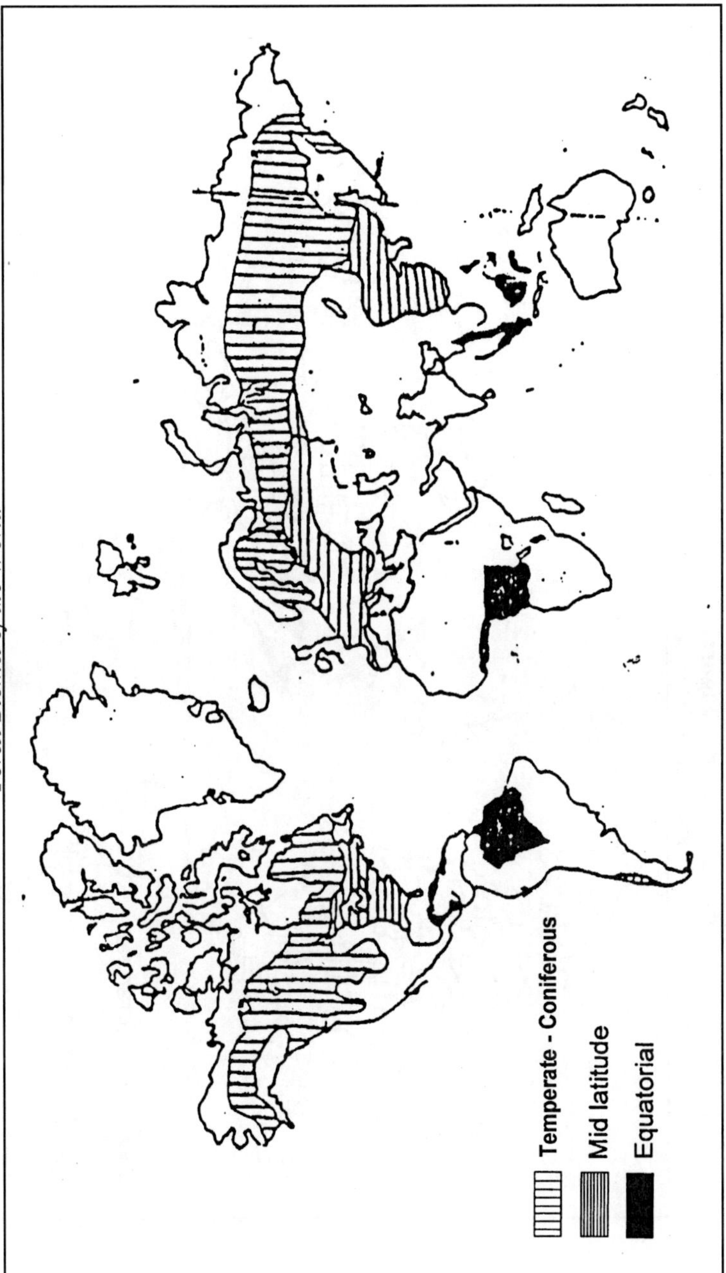

Figure 2.11
Forest Biomes of the World

Figure 2.12

Grassland and Mediterranean Biomes

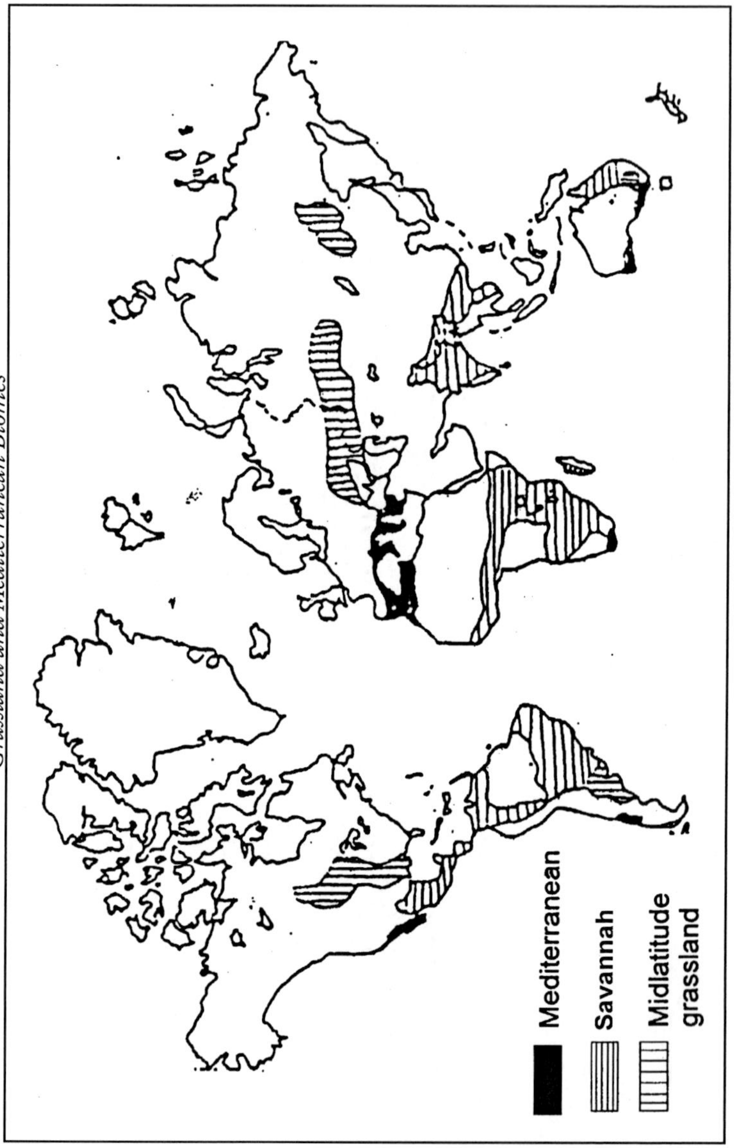

Figure 2.13

Barren Areas Biomes

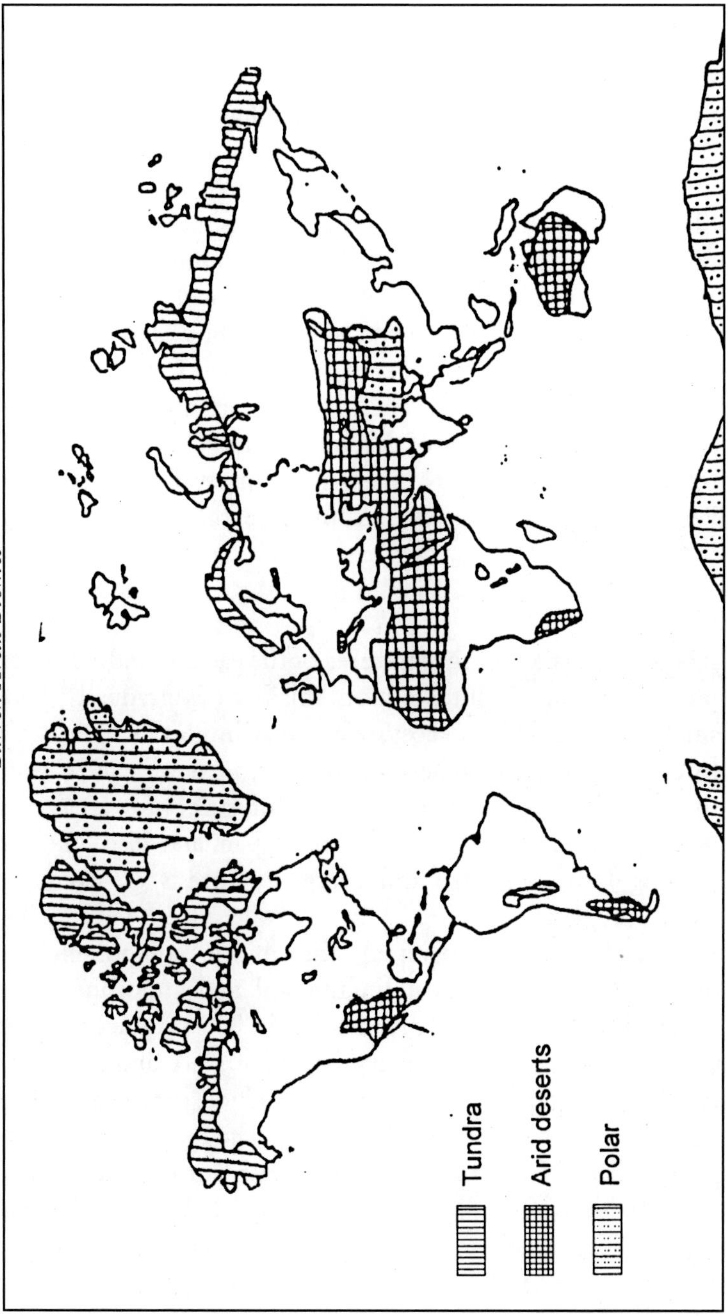

Grassland Ecosystem

Grasslands occupy about 19 per cent of the earth's area, which include tropical and temperate grasslands. In this, the savannah ecosystem is very important. The abiotic components are the nutrients present in the soil and aerial environment. The elements like carbon dioxide, water, nitrates, phosphates, sulfates, etc., are present in the air and soil of the area. The producers are mainly grasses and small trees and shrubs. The primary consumers include cows, buffaloes, sheep, goats, deer, rabbits, and other animals, while secondary consumers are animals like foxes, jackals, snakes, frogs, lizards, birds, etc. The microbes are active in the decaying and dead organic matter of different forms. They bring the minerals back to the soil, thus making them available to producers. Pastoralism and livestock ranching are the main occupations in these regions.

Forest Ecosystem

About 30 per cent of the land area of the earth is under forest cover, but due to man's intervention this area is gradually becoming smaller. But still forest ecosystem is very important. The coniferous forests stretch as broad belts across North America and Eurasia. On the other hand, temperate deciduous forests occupy areas in eastern North America, parts of Europe, Japan and also in Australia. Among forests the tropical evergreen forests are found in the tropical regions.

In the forest ecosystem, the abiotic components are the organic substances present in the soil and atmosphere and also minerals present in dead organic debris. The producers are trees of different species. The primary consumers are animals of various types, while secondary consumers include carnivores like lions, tigers, snakes, birds, lizards, foxes, etc. The forest ecosystems are of great concern from the environmental point of view. The rate of exploitation of forest by man is growing day by day, thus causing a great concern to all nations of the world because of its impact over global climate and on several animal species.

Desert Ecosystem

Deserts generally occur in regions having less than 25 cms of rainfall and are unevenly distributed. Scarcity of rainfall may be due to: (i) high sub-tropical pressure, as in the Sahara and Australian deserts, (ii) geographical position in rain shadows, or (iii) high altitude as in Tibetan, Gobi, Bolivia deserts. There are three life forms of plants that are adapted to deserts: (i) the annuals, which avoid drought by growing only when there is adequate moisture, (ii) the succulents, such as the cacti, which store water, and (iii) the desert shrubs, which have numerous branches and a special root system which help them to adapt to desertic conditions. The most common animals, apart from camels and goats, are reptiles and insects, able to live under xeric conditions. Due to poor vegetation, dead organic matter is less and some fungi and bacteria act as decomposers. Apart from hot desert, there are also widespread cold deserts encircling the north and south poles, where permanent ice caps have developed an entirely different type of ecosystem, sustainable to cold climatic conditions.

Cropland Ecosystem

Apart from above mentioned natural ecosystems, there are also man-engineered ecosystems. One such ecosystem is the cropland ecosystem, in which man has developed croplands after considering the soil, climatic and other environmental conditions. These are ecosystems of dominant crop species like wheat, maize, jowar, paddy, sugarcane, cotton, tea, coffee, etc.

The abiotic components of this ecosystem include climatic conditions and mineral contents of the soil. In case of any deficiency man used chemical fertilizers and/or water for irrigation, etc. The various types of foodgrains, pulses and commercial crops are grown in these croplands, which provide food and fodder not only to man and animals, respectively, but in the fields several types of animals like birds, rats, rabbits, and other smaller reptiles insects, etc., also survive. The decomposition of dead organic matter of plants and animals make the minerals available again.

There is a need that every ecosystem should be studied in detail. Apart from above mentioned macro or global ecosystems, there are several meso and micro regions existing in each region. For example, India has the following major ecosystems:

1. Himalayan mountain ecosystem
2. Plain ecosystem
3. Desert ecosystem
4. Central India plateau ecosystem
5. Peninsular (Deccan plateau) ecosystem
6. Coastal plain ecosystem
7. Island ecosystem

These ecosystems have been identified on the basis of physiography, which also controls climate, natural vegetation, soil and also man's adoption to nature. Each major region can be divided into sub-ecosystem regions; for example, Rajasthan has the following ecosystems:

(i) Arid ecosystem
(ii) Irrigated arid ecosystem
(iii) Sub-arid ecosystem
(iv) Aravalli hills ecosystem
(v) Eastern plains ecosystem
(vi) Hadoti plateau ecosystem

Further identification of micro-ecosystems will help not only in understanding the ecosystem but also in regional development and planning. The goal of sustainable development can only be achieved, when our developmental policies are based on the micro-ecosystem of each region.

Notes

1. Monkhouse, F.J. and Small, J., *A Dictionary of Natural Environment*, 1980 p. 38.
2. Haggett, P., *Geography : A Modern Synthesis*, 1975, London: Harper and Row, p. 111.

3. Strahler, A.N. and Strahler, A.H., *Geography and Man's Environment*, 1977, p. 271.
4. *Encyclopaedia Britannica*, Vol. III, 1980, p. 780.
5. Odum, E.P., *Fundamentals of Ecology*, 1971, p. 43.
6. Ibid., p. 86.
7. Odum, E.P., op. cit., 1971 p. 378.
8. Ibid., p. 295.
9. Ibid., p. 324.

3

Human Impact on Environment

The human impact on natural environment is an important component in the management and optimisation of the *man-society-nature* system. It is also one of the most pressing issues of contemporary times and a subject of discussion in geographical literature as well. The way human beings have changed and are changing the face of the earth has drawn the attention of not only natural scientists but also social scientists as well as planners and policy makers. The basic law of ecology is that everything is connected to everything else and that one cannot change just one thing in nature. George Perkins Marsh published *Man and Nature* in 1864, in which he dealt with human influence on woods, waters and sands. This relationship has also been described in depth in their own way by geographers like Brunhes, Blache, Semple, Ratzel, Barrows, Sauer, Gerasimov, etc. The subject of the human impact on environment, dealing as it does with such matters as environmental degradation, pollution and desertification, etc., has also been discussed in many textbooks and research monographs. Although the entire text of this book is in some way or other related with this aspect, the present chapter aims to discuss:

(i) human impact on natural environment, and
(ii) degradation of environment due to human activities.

Human Impact on Natural Environment

Human activities have had a great impact on natural environment. This impact can be seen on climate and other atmosphric phenomena, on vegetation, soils, animals, water as well as on geomorphological processes. A brief review of the human impact on natural environment is given here just as an introduction which provides a base for environmental management and planning.

Climate and Atmosphere

The increasing human population and the advancement in technology have not only become significant factors in the variations in world climate but are also responsible for the various changes in atmospheric conditions including air pollution. The human influence on global climate is due to the following mechanisms:

- gas emissions
- CO_2—industrial and agricultural
- methane
- chlorofluorocarbons (CFCs)
- nitrous oxide
- krypton 85
- water vapour
- miscellaneous trace gases
- aerosol generation
- thermal pollution
- albedo change
- dust addition to ice caps
- deforestation
- overgrazing
- extension of irrigation
- alteration of ocean currents by constricting straits
- diversion of fresh waters into oceans

The problem of carbon dioxide (CO_2) emission has become a major environmental concern. Since the beginning of the Industrial Revolution humans have been taking stored carbon out of the earth in the form of coal, petroleum and natural gas, and burning it to make carbon dioxide, heat, water vapour and small amounts of sulfur dioxide (SO_2) and other gases, which are responsible for air pollution, green house effect, increase in surface temperature, or in other words, global warming. By 2050, it is possible that the increase in global surface temperatures ranges between 1.4 and 2.2 degree celsius. All other aspects of atmospheric pollution will be discussed under the heading, 'Air Pollution'.

The impact of human activities on the atmosphere is more because the atmosphere acts as a major channel for the transfer of pollutants from one place to another. It is in this way that harmful substances are transferred long distances from their sources of emission. Another example of the possible widespread and ramifying ecological consequences of atmospheric pollution is provided by acid rain.

In recent years, the greatest attention has been paid to the role of CFCs, the production of which has been rising in last few decades. These gases may diffuse upwards into the stratosphere where solar radiation causes them to become dissociated to yield chlorine atoms which react with and destroy the ozone present there. The Antarctic ozone hole has been identified through satellite monitoring. In fact, air pollution and its variability both in space and time is more due to human action.

Vegetation

The human impact on vegetation is greater than on any of the other components of the environment. The nature of whole landscape has been transformed by human-induced vegetation change as depicted in Figure 3.1.

The extent of human influence on vegetation is provided by Westhoff (1983) in the following four-part scheme:

(i) *Natural*—a landscape or an ecosystem not influenced by human activity.
(ii) *Sub-natural*—a landscape or ecosystem partly influenced by humans, but still belonging to the same formation type as the natural system from which it derives.
(iii) *Semi-natural*—a landscape or ecosystem in which flora and fauna are largely spontaneous, but vegetation structure is altered so that it belongs to other formation type such as pasture, moorland, etc.
(iv) *Cultural*—a landscape or ecosystem in which flora and fauna have been essentially affected by human agencies in such a way

that dominant species may have replaced by other species as in case of arable land.

Figure 3.1

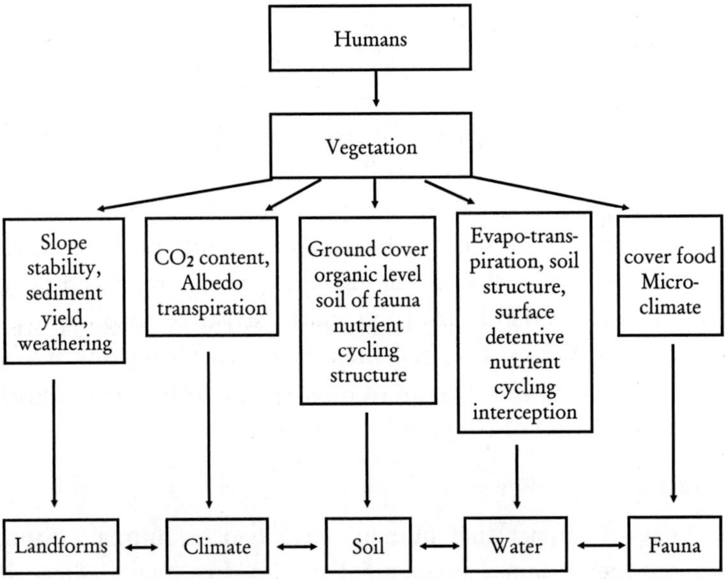

Man has used fire for the clearance of forest cover, mainly for the use of land for cultivation or for habitation. This practice was common during the early stages of the development of civilisation. But, it is still prevalent among many tribals. Although controlled burning is a part of forest management, uncontrolled and deliberate fire destructs not only vegetation but also animals and other forest creatures. Similarly, uncontrolled and heavy grazing is not only a cause of the disappearance of vegetation cover but is also responsible for desertification and other environmental problems.

The deliberate removal of forests or deforestation, is one of the most long-standing and significant ways in which humans have modified the environment. Sometimes forests are cut down to allow agriculture, at other times to provide fuel for domestic purposes or to provide charcoal or wood for construction, etc. One of the great phases of forest clearance occurring at the present time is in humid tropics. The annual regression rate of tropical moist forest is about 11 million hectares. The spread of desert-like conditions (desertification) in arid and semi-arid areas is always due to man's action. Some of the air pollutants, which humans have released into the atmosphere, have had detrimental effect on plants, for example, sulfur dioxide, which is toxic to them. Local concentrations of industrial fumes also kill vegetation. Photochemical smog is also known to have adverse effect on plants both within cities and also on their outskirts. The adverse effects of pollution on plants are not restricted to air pollution; for water and soil, pollution can also be serious. Excessive amounts of heavy metals may prove toxic to them. Salt marshes, mangrove swamps and other kinds of wetlands are particularly sensitive to oil spills. In fact, human activities have become a major cause not only for the gradual removal of forest cover in many parts of the world but also for the degradation of the environment.

Animals

The range of impact that humans have had on animals, though large, can be grouped conveniently into five main categories: domestication, dispersal, extinction, expansion and contraction. The extinction of animals by human predators has been extensive over the past 20,000 years, and in spite of recent interest in conservation, it continues at a high rate.

The extreme effect of human interference with animals is extinction. The problem of extinction of several animal species has become a world-wide ecological problem. This decline is partly due to intentional killings for subsistence and commercial purposes, but much wildlife decline occurs due to pollution and also due to change in ecosystem and natural habitat. For example, the use of DDT is one of the causes for the extinction of several types of insects, birds and small animals. The polluted water containing heavy metals and methyl mercury is also harmful. Oil pollution is an increasingly serious problem for marine and coastal fauna and flora. Sea birds are especially vulnerable since oil clogs their feathers. Other industrial pollutants have a clear impact on aquatic systems. Pollution is a serious problem for coral reefs too.

The list of other indirect causes of wildlife decline is a vast one. Modern society is particular about the conservation of wildlife and efforts have been made by international, national and NGOs to protect them, but due to selfish and commercial motives, the killing of these animals is still going on. Various types of habitats have also started disappearing due to adverse human impact. On the global basis the loss of wetland habitats is a matter of considerable concern. Such areas are important for the breeding of several animals. The other threats include drainage, dredging, mining, filling, pollution and channelisation.

Briefly, the human impact on animal extinction can be discussed in three categories as under:

(i) the *blitzkrieg effect*, which involves rapid development of human populations with big-game hunting technology so that there is very rapid demise of animal population;
(ii) the *innovation effect*, whereby long-established human population groups adopt new hunting technologies and erase fauna that have already been stressed by climatic changes; and
(iii) the *attrition effect*, whereby extinction takes place relatively slowly after a long history of human activity because of loss of habitat and competition for resources.

Soils

Soil is the most vulnerable of human resources and is one on which humans have had a very major impact, because they live close to and depend on it. Impact on soil can occur with great rapidity in response to land use change by new technologies. The major changes brought about by humans are chemical, structural and hydrological, while soil erosion is perhaps the most important.

Salinity is a natural characteristic in some semi-arid and arid soils. But humans have increased the extent and degree of salinity in different ways. The extension of irrigation and different techniques, used for water abstraction, can lead to a build-up of salt levels in the soil through the mechanism of raising groundwater level. The construction of large dams and barrages to control water flow and to give a head of water creates large reservoirs from which further evaporation can take place. The seepage of water is also responsible for upward movement of groundwater. In coastal areas, salinity problems are created by sea water incursion brought about by overpumping. The clearance of native forest vegetation is also a cause of increase in salinity. Measures have been taken for reclamation of salt affected areas through eradication, conversion and control techniques, but soil salinity is still a grave problem in many parts of the world.

Laterisation is a chemical process in the soil of the tropics where extensive sheets of a material called laterite have been deposited either through natural process or by human actions. Laterite is a paleological leprosy. Man's activities aggravate the dangers of laterisation when started by negligent removal of the forest.

Human activities are also responsible for the structural changes in soil. There are many ways in which humans can alter this, especially by compacting it with agricultural machinery and by changing its chemical character through irrigation. Grazing is another activity that can damage soil structure through trampling and compaction. Soil drainage is a slow and gradual process and is responsible for certain environmental changes and also a cause for long-term damage to soil quality.

The introduction of chemical fertilizers has also changed the chemistry of soils. Sometimes, these create environmental problems such as water pollution, while their substitution for more traditional fertilizers may accelerate soil structure deterioration and soil erosion. Nowadays, the use of three main fertilizers, viz., nitrates, phosphates and potash, has greatly increased the agricultural productivity but it has also changed the basic structure of the soil.

Soil erosion is the result of natural processes but at the same time it is also a result of human activities. Myers (1988) has summarised the scale of accelerated soil erosion that has been done by human actions: "Since the development of agriculture some 12,000 years ago, soil erosion is said by some to have ruined 4.3 million

km^2 of agricultural lands, or an area equivalent to rather more than one-third of today's croplands—the amount of agricultural land now being lost through soil erosion, in conjunction with other forms of degradation, can already be put at a minimum of 200,000 km^2 per year." In fact, soil erosion is a major and serious aspect of the human role in environmental change. The processes of construction, urbanisation, mining, etc., are also accelerating this problem, although the prime causes of soil erosion are deforestation and agriculture. Because of the adverse effects of accelerated erosion, several techniques have also been used to conserve soil resources. Some of them are revegetation, stoppage of full enlargement, crop management, slope run-off control, suppression of wind erosion, terracing, contour ploughing, etc.

Water

Water is the source of life and, human beings use it for various purposes. The ancient civilisations have developed in river valleys. This is also true of medieval townships and other developments, and all modern developments are related directly or indirectly to water. The main concern is that by using water, humans have influenced both its quantity and quality. Earlier, the influence of human activity on water resources was limited but now this has become a major problem of environmental degradation throughout the world. However, there are many ways in which humans influence water, for example, by direct channel manipulation, modification of watershed characteristics, urbanisation and pollution.

The construction of dams and reservoirs is widespread throughout the world for irrigation, to generate power or to provide a reliable source of water. More than 700 dams have been built worldwide and their number is increasing year after year. In countries like Brazil, Argentina, Canada, India, Japan, Turkey, Spain and China, run-off of water has been regulated by water reservoirs. The impact of human activities is especially marked in Africa and North America where about 20 per cent of total run-off is now controlled. The impact of dam, construction is apparent in the form of change in ecosystem and environmental conditions. Bigger the dam more will be its impact. Dams like Hoover (USA),

Kiev (Russia), Aswan (Egypt), Akosombo (Ghana), and Bhakra Nangal, Gandhi Sagar, Damodar and Rihand (India) have had their impact on regional ecology. The recent environmental controversy over Tehri and Sardar Sarovar dams shows people's awareness against construction of dams and their possible damaging environmental impact. The environmental consequences of dams include subsidence, earthquake triggering, transmission and expansion in the range of organism, the build-up of soil salinity, changes in groundwater levels, waterlogging, deforestation, etc.

The process of urbanisation has a considerable impact, in terms of controlling rates of erosion, the delivery of pollutants to rivers and influencing the nature of run-off and other hydrological characteristics. The impact of urbanisation can be summarised as follows:

(i) removal of trees or vegetation,
(ii) construction of houses and sewerage lines,
(iii) construction of septic tanks,
(iv) diversion of nearby streams for public supply,
(v) accelerated land erosion,
(vi) larger quantities of untreated waste into local streams,
(vii) pollution of streams and wells,
(viii) drilling of deeper, large capacity industrial wells, and
(ix) increased use of water for domestic and commercial purposes, etc.

Deforestation is a human activity that gives rise to floods, annual run-off levels and also affects on-stream flow. In many parts of the world humans obtain water supplies by pumping from ground water. This has two main effects, i.e., the reduction in the level of water table and the replacement of coastal areas of fresh water by salt water. Environmental consequences of these two phenomena include ground subsidence and soil salinisation.

Water pollution has now become a common problem and is a result of human use and misuse of water for various purposes. The causes and forms of water pollution created by humans are many and can be classified into the following groups:

1. Sewage and other oxygen-demanding wastes
2. Infectious agents
3. Organic chemicals
4. Other chemical and mineral substances
5. Sediments
6. Radioactive substances
7. Heat (thermal pollution)

Many human activities can contribute to changes in water quality including agriculture, fire, urbanisation, industry, mining, irrigation, etc.

Geomorphology

Apart from above mentioned impact of human activities on the natural environment, the human role in creating landforms and modifying the operation of geomorphological processes such as weathering, erosion and deposition also needs consideration. The range of the human impact on both forms and processes is considerable. Haigh (1978) has classified anthropogenic landforming processes as follows:

(a) *Direct Anthropogenic Processes*
  - (i) Constructional: tipping–loose, compacted, molten; graded–moulded, ploughed, terraced.
  - (ii) Excavational: digging, cutting, mining, blasting, etc.
  - (iii) Hydrological: flooding, damming, canal construction, dredging, channel modification, draining, coastal protection.

(b) *Indirect Anthropogenic Processes*
  - (i) Acceleration of erosion and sedimentation, agricultural activity, clearances of vegetation, engineering, road construction, urbanisation.
  - (ii) Subsidence: collapse, settling, mining, hydraulic, thermokarst.
  - (iii) Slope failure: landslide, flow, accelerated creep, loading, undercutting, etc.
  - (iv) Earthquake generation: loading (reservoirs) lubrication (fault plane).

Human Activities and Environmental Degradation

In the present century, environmental degradation has emerged as a major global concern for human survival. In the opinion of the World Commission on Environment and Development (WCED), the future generations are to face an ever-increasing environmental decay, poverty, hardship and a more polluted world. The strains of the ecological crisis are so apparent that the task to preserve and protect the environment has become the primary requisite of the economics of development. Modernisation is steadily transforming the nature into products. The resource squeeze has led to an ever-intensification of ecological damage. Land, water and air are being polluted every minute. Approximately, 10,000 million tonnes of oxygen is being burnt and about 24,000 million tonnes of carbon dioxide is being discharged every year. The green revolution has turned red. Deforestation is proceeding an alarming rate depleting the genetic store. Mega technology has created problems of waste disposal and industrial revolution has now become a great hazard to the environment as well as to humanity.

Man's environment by now is sufficiently saturated by the complex chemical emissions, aerosols, toxic effluents, sewage, pesticides, solid wastes, polluted rains, dust and radiation. UN report on Global Environments reveals that the globe will become more crowded, more polluted, ecologically less stable and more vulnerable to disruption than the world we live in today. Degradation of environment has now become one of the most challenging problems not only for developing countries but also for the developed world. A large number of international organisations with the collective wisdom of scientists, planners and geographers have come up to settle the environmental issue confronting the most nations. The environmental crisis has convinced the world to use technology and resources to repair the damage already done to our environment and also to use substitutes for certain harmful chemicals in order to protect and preserve nature and natural resources.

The environment is an integrated system in which all its elements act and react in such a way that a balance is always maintained. All the physical elements such as relief, structure, climate, natural vegetation, soil, water bodies, etc., are the determining factors of environment on the one hand, and man's activities on the other. Man is a user of the environment for his developmental activities and always disrupts this natural system and creates a background for environmental degradation. When this degradation is lesser, nature recovers it by its own system of recovery, but whenever its quantity is more, it creates an imbalance in nature.

According to Paul Harrison, "rape of the earth has been central to the past progress of western civilisation. The plunder and waste of irreparable resources, the poisoning and destruction of the fragile ecosystem, the belief that man can mould the nature with no need to take account of nature's reaction to such violation is a gross oversight". Scientific and technological developments have given several benefits to mankind but at the same time, they are also responsible for various types of environmental degradation. The impact of human activities on natural environment has already been discussed, but its impact on resource utilisation and other economic and social elements has also been responsible for the degradation of environment. The various forms of pollution are the result of human activities, which will be discussed in other chapters. Here our discussion is limited to deforestation, construction of big dams, mining, use of pesticides, urbanisation, industrialisation, etc., which are responsible for the degradation of environment.

Deforestation

The herculean feats of economic gains and development have been at the cost of deforestation. Deforestation is one of the outcomes of various developmental activities having the character of extensional development and it is true that deforestation is a havoc wreaked by thoughtless destructive activities of development. Forest clearings were made not only for village settlements but also for cultivation and pastures. As the population increased, more forests were cleared for various uses. Apart from this, the commercial exploitation of forests is the main cause of deforestation. There was a time when 70 per cent of the land area was covered with forests and now the total forest cover has shrunk to 16 per cent only. Nowadays, tropical deforestation has become a concern for all and its impact is not only limited to the region itself, but on the whole world. The annual decline in forest cover in tropical countries has been depicted in Figure 3.2. It is clear from the diagram that in countries like Brazil, India, Indonesia, Thailand, etc., the rate of annual cutting of forests is very high and has become a great problem to the environment and ecosystem.

In India, deforestation is going on at a very fast rate. The story of overexploitation of forests started from World War II, when the British started clearing forests to meet their needs. In the post-independence period, the process has been accelerated to meet developmental needs of railways, industries, mining, river valley projects, defence, etc. The urbanisation and expansion of agricultural activity have also contributed to the exploitation of forests. During the 1980–2000 period, 28.67 lakh hectares of forest land have been lost to different developmental works. The commercial exploitation of forests is still going on in spite of several legal provisions to restrict it. Maximum deforestation has occurred in Madhya Pradesh, where loss of forest is about 2 million hectares, while in Maharashtra, Orrisa, Andhra Pradesh, Jammu & Kashmir over a million hectares and Rajasthan and Himachal Pradesh over half a million hectares have been cleared. But, in terms of proportion, Rajasthan, Haryana and Punjab lost over half of the forest cover and states like Himachal Pradesh and Jammu & Kashmir lost over a third of their forest cover. In Tripura, Manipur, Meghalaya and Mizoram the loss of forest cover was minimum during the 1970–2000 period.

The impact of deforestation on the environment can be seen in the form of micro-climatic change, increase in temperature and decrease in humidity/rainfall, soil erosion in the form of sheet, rill and gully erosion, increase in frequency of floods and landslides, loss of soil fertility, etc. All these lead to unaccountable destruction and also pose serious ecological challenges. Serious efforts will have to be made to protect and conserve forests both at government as well as public levels. Due attention should also be given to forest management. Some useful steps in this direction are:

Figure 3.2

Area of Closed Forests Cleared Annually in Selected Tropical Countries

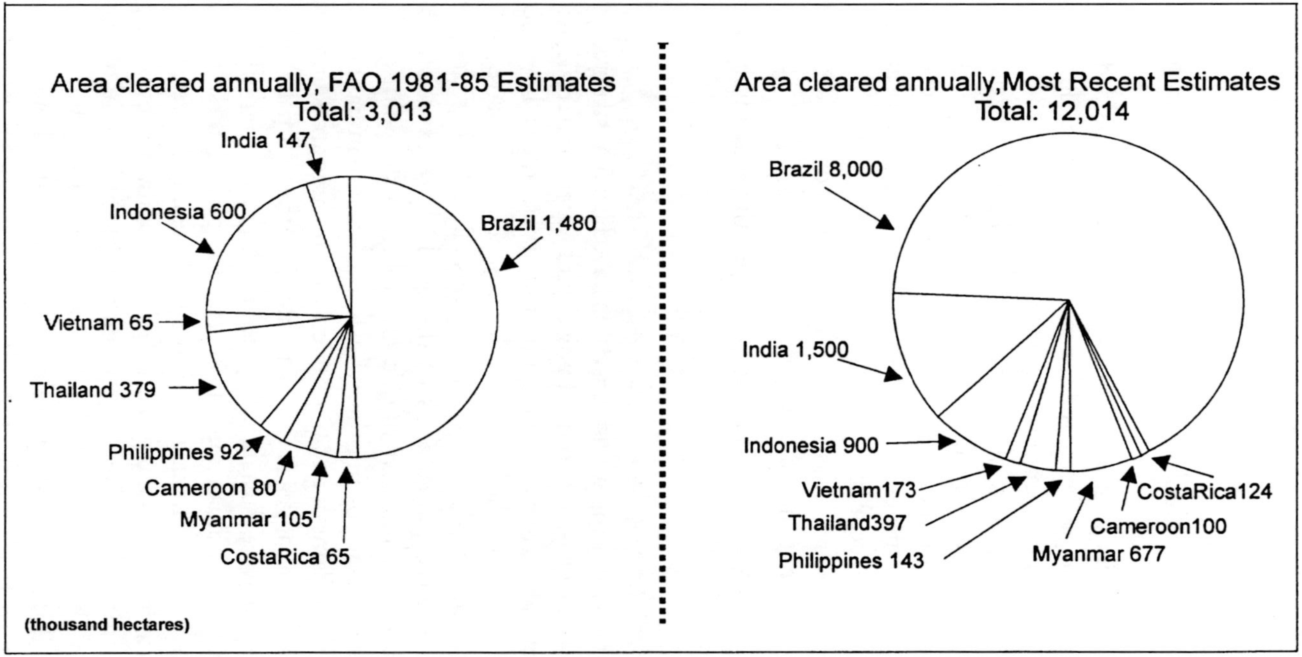

(i) a proper ratio between forest cutting and plantation of trees should be maintained;
(ii) actual growth of trees should be supervised;
(iii) strict prohibition of the cutting of trees for fuel should be imposed;
(iv) alternative media for wood such as moulded plastic, etc., should be developed for furniture and other domestic articles;
(v) protection from forest fires;
(vi) development of national parks and game sanctuaries;
(vii) social forestry should be developed;
(viii) plantation of trees according to local ecological conditions;
(ix) forest surveys should be conducted and classification should be done accordingly; and
(x) long-term master plan for forest conservation and development should be prepared.

In fact, forests can be protected through public movements like '*Chipko*'.

Big Dams

Throughout the world, big dams have been constructed in order to fulfil the growing needs of power and water for irrigation and other purposes. There is no doubt that these dams have become an indicator of the level of development and have boosted the economy of the respective regions. But, during the last twenty years, environmentalists of the world have expressed the feeling that these dams may not be suitable from the point of view of the environment, or in other words, the ecosystem has been disturbed by these dams due to (i) submergence of large areas under water, (ii) destruction of vegetation, wildlife and other animals, (iii) instability may provide sustainability to earthquakes, (iv) silting, and (v) large-scale human migration, etc. In spite of these problems, the construction of dams continues unabated. However, it is necessary to examine big dam projects from the point of view of degradation of environment and then only implementation of the project should be allowed. As far as possible, emphasis should be given on small dams. Similarly, selection of dam site should be done in a manner that there is minimum impact on the environment.

In India, during post-independence period, several dams—both small and big—have been constructed. Certainly, these have changed the entire scenario of power supply and agricultural development. However, despite their negative impact on environment, all these projects are considered beneficial for the people. The controversy over the Tehri and Sardar Sarovar projects is the example of negative impact of big dams.

The Tehri project has gained significance mainly because of the dimensions of the dam. The points raised by environmentalists are:

(i) The ecology will suffer, invaluable forests will be destroyed because of submergence of forest lands. The wildlife and birds, etc., living in these areas, would face extinction.
(ii) The possibility of earthquakes would be more in view of the large size of the dam.
(iii) The deposition of silt in the base of dam may restrict flow of huge quantum of water.
(iv) The stagnant waters of the reservoir would spread water-borne diseases in adjoining areas.
(v) The rehabilitation of tribals and other people may not be proper.

All these issues are raised for other such projects also and experts are of the opinion that for sustainable development it is necessary to build only small dams. These will not only save cost and energy but will also be harmless for the environment. However, whenever there is national necessity, the decision should be taken after considering the environmental issues also.

In the case of Tehri dam, apprehension about its safety has been expressed by many on account of the fact that the dam is going to be located in a region of high seismocity and is, therefore, potentially unsafe. This has, however, been denied by experts. There is no doubt about the fact that the Tehri dam will change the entire ecosystem of the region, but looking at the benefits in the form of power generation and water supply, this project has not only been approved by the government but its work is near completion. In fact, issues raised regarding environmental degradation are present, but what is not generally pointed out is that environmental degradation is already there.

Similarly, the Sardar Sarovar project has been designed to provide irrigation water to drought-hit areas of Gujarat, Madhya Pradesh, Rajasthan and Maharashtra. This project, which is estimated to cost over Rs 6,400 crores, will on completion, provide irrigation facilities to about 18 lakh hectares of land and create installed capacity of power of 1,450 MW. It will benefit about 25 lakh people in 3,340 villages. It will also provide drinking water for 295 lakh people residing in 131 towns and 4,720 villages in Gujarat. But it will create serious environmental problems, for which proper steps should be taken in advance in order to maintain the regional eco-balance.

Mining

Mining is one of the most important economic activities the world over. With growing knowledge of the mineral resources as well as due to technological and scientific developments, mining is today performed on a very large scale. The extraction of minerals is done not only to fulfil regional or national needs but also for trade. With the result the large-scale mining operations have become a threat to the environment.

Mineral resources are the product of geological structure and their formation may take a very protracted period, extending over several thousands years. Therefore, any mineral extracted cannot be formed again for use. There are three types of mineral resources, viz., non-metallic minerals, metallic minerals and mineral fuels. All these are useful to man in some way or the other. Therefore, mining operations are expanding day by day without taking into consideration its adverse impact on the environment. Digging of land is carried out on a very large scale and impacts the environment in multiple ways:

(i) In mining areas destruction of land due to deep and surface mining is a common feature. In USA alone, about 1,50,000 acres of land has become wasteland due to mining. In areas of coal, iron ore, building material and other ores, etc., one sees no signs of the original relief. In mining areas of Bihar, Orrisa,

Madhya Pradesh and Rajasthan, one can already see the destruction of land.

(ii) Heaps of unused material is also a great problem. Generally mine owners do not care for this waste material, therefore, it becomes the main cause of environmental degradation.

(iii) Large-scale destruction of natural vegetation and wildlife creates ecological problems.

(iv) Pollution due to dust spreading all over the area is not only a threat to the environment but also the cause of many diseases among the workers as well as the residents of nearby areas.

The other impacts may include:

(a) expansion of barren land,

(b) creation of 'ghost town' after mining, and

(c) ecological crisis due to air, water, land and noise pollution.

Unplanned and unorganised mining is a danger to the environment and this should be checked. Measures in this direction could be:

(i) levelling of mining areas,

(ii) re-use of waste material,

(iii) plantation in mining areas,

(iv) rehabilitation of people,

(v) measures to check air and land pollution,

(vi) improvement in mining areas,

(vii) health and safety measures,

(viii) policy of resource conservation should be adopted and alternative material should be developed, and

(ix) proper management and planned regulated mining can solve many problems.

Pesticides

The problem of contamination through pesticides in our food grains, dairy products, vegetables, fruits or in our living environment as a whole has been visualised very well. Pesticides have been used in farming (i) to minimise the loss of food caused by pests and ensure a more bountiful harvest, and (ii) to decrease the extent of vector-borne diseases. Contrary to these benefits, it has seriously degraded environmental quality and altered the ecological balance, as shown in Figure 3.3.

Figure 3.3
Pesticide Impact Cycle

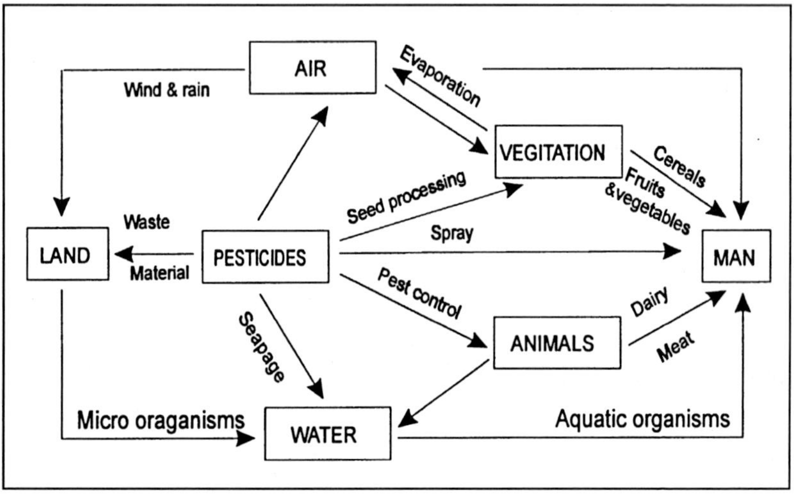

The pesticides, which have been widely applied throughout the world, are carried forth by wind and water and distributed elsewhere. It was found that the some pesticides slowly decomposed into compounds that were toxic to fish and wildlife species in recent years. Worldwide, 70,000 chemicals are in use per day and 500 to 1,000 new ones are added each year. Between 4,00,000 to two million pesticide poisonings occur worldwide each year, of which an estimated 10,000 are fatal.

Research has proved that continuous use of pesticides has many negative effects on the environment, such as: (i) killing non-targeted organisms, (ii) accumulation in the food chain, (iii) building of immunities by the targeted pests, (iv) lower reproductive potential, (v) synergistic effects, (vi) toxic effects on water, (vii) pesticide residue in food and tissues, (viii) effects on vegetation and milk, and (ix) direct pesticide poisoning.

Pesticides have their impact on all three components of the earth, i.e., lithosphere, hydrosphere and atmosphere.

Soil: Pesticides are toxic, hence it is assumed that their presence in soil will change or alter fundamental soil processes, such as: (a) organic matter decomposition, (b) nitrogen transformation, (c) sulfur transformation, (d) phosphate availability, (e) trace element availability, and (f) soil enzyme activity, which influence soil fertility and productivity.

Water: The presence of toxic chemicals in water also has a significance as they are picked up by unicellular aquatic organisms like plankton and get accumulated in the body by a phenomenon called bioconcentration.

Air: The amount of pesticides released into the atmosphere, if larger than really required, pollutes the air.

Presently, in India, 123 pesticides are permitted and are being used over a total area of about 142 million hectares. Among all pesticides, insecticides constitute about 80 per cent, followed by fungicides at 10 per cent, herbicides at 7 per cent, and others at 3 per cent. Pesticides like DDT, DDE, TDE, deidrin, endrin, heptachlor-epoxide and aldrin are harmful. Many developed countries have prohibited the use of those pesticides which have harmful effects on living organisms. Pesticides can even enter the body through the skin, lungs (inhaled) or the digestive tract. Nausea and vomiting, abdominal cramps, diarrhoea and involuntary defecation are the usual symptoms following gastro-intestinal poisoning.

Many countries have enacted laws for the safe use of pesticides. In India, there are two laws, i.e., Prevention of Food Adulteration Act, 1954 and the Insecticides Act, 1968. In fact, the use of pesticides should be restricted and pest control should be done by other methods and also quality of pesticides should always be supervised and maintained.

Environmental Degradation and Technological Development

With the help of scientific and technological developments man has achieved new dimensions in his own development and has improved his quality of life by adding several amenities and facilities. The industrial development is proceeding at a very fast rate; similarly, revolution in the transportation sector has also occurred not only in developed countries but also in developing countries. In the fields of agriculture, resource utilisation, power consumption, engineering and in other scientific fields man has achieved success of high level. But, the current strong feeling is that all these developments have been responsible for the degradation of the environment and there is an urgent need to evolve an environmental-friendly technology. It is said that "technology produces the crisis and technology can solve it". Some of the problems produced by technological and scientific developments are as follows:

(i) Degradation of environment and expansion of pollution through industrialisation.
(ii) Unorganised mining.
(iii) Rapid deforestation.
(iv) Pollution due to transportation.
(v) Danger of leakage of radio-activity from atomic reactors.
(vi) Disturbances in eco-sensitive areas like ozone layer, tropical forests, Antarctica, etc.
(vii) Population growth, urbanisation, decline in death rate and other demographic changes are responsible for more and more exploitation of resources.
(viii) Extinction of several species of wildlife, birds and plants.
(ix) Too much stress and strain has created several health problems.

In brief, the quality of environment is deteriorating day by day. It should be maintained through environment planning, people's participation, environmental status evaluation, and environmental legislation and administration.

Conclusion

The above mentioned impacts of human activity on environment are just an introduction to the thinking that with the development in technology, the number of ways in which humans are affecting the environment are proliferating. The complexity, frequency and magnitude of impacts is increasing, partly because of rising population and party because of a general increase in per capita consumption. The US Council on Environmental Quality in 1982 had predicted the conditions of environmental degradation for year 2000, which are still relevant. The report states:

> If present trends continue, the world in 2000 will be more crowded, more polluted, less stable ecologically, and more vulnerable to disruption than the world we live in now. Serious stresses involving population, resources and environment are clearly visible ahead. Barring revolutionary advances in technology, life for most people on earth will be more precarious in 2000 than it is now unless the nations of the world act decisively to alter current trends.

In fact, nations of the world are coming closer at least on environmental issues and they are thinking in terms of sustainable development. The technology has produced several environmental problems and technology can only solve these problems.

References

Haigh, M.J. (1978), Evolution of Slopes on Artificial Landforms, Blaenavon, University of Chicago, Research Paper 183.

Northam, R.M. (1975), *Urban Geography,* John Wiley, p. 76.

Mann, H.S. (1977), Desertilization: Overview, *Arid Zone,* 16 (3), pp. 1–4.

Myers, N. (1988), *Natural Resource Systems and Human Exploitation Systems: Physiobiotic and Ecological LInkages.* World Bank Policy Planning and Research Staff, Environment Department Working Paper, 12.

Weshoff, V., (1983), "Man's Attitude Towards Vegetation" in W. Holznex, M.J., A. Werger and I. Kushima (eds.), *Man's Impact on Vegetation.* The Hague: Junk, 7-24.

4

Management of Environment: Approaches and Measures

In the last decade of the 20th century, environmental degradation has emerged as a major global concern for human survival. A large number of international and global organisations, with the collective wisdom of scientists, economists and planners, have come up to settle the environmental issues confronting the nations—both developed and developing world. The environmental crisis has convinced the world to use technology and resources to repair the damage already done to our environment and also to patent substitutes for certain harmful chemicals in order to protect and preserve nature and natural resources. The last three decades have witnessed tremendous awareness in the field of environment all over the world. The Dictionary of Environment is renewed every time with new terminologies like clean technologies, environmental auditing, environmental-friendly products, etc. Similarly, the concept of environmental management has also emerged and accepted as a tool for sustainable development. In fact, correct use of technology, natural resources and human resources can be done through their proper management. Many problems related with environment and its degradation can be solved with the help of its proper management. The developed countries of the world have already developed a system of environmental management at all levels, i.e., national, provincial and local. But, in developing countries, this has yet to be done through legislation and people's participation.

If the condition of the global environment is to be improved, we must clearly define the ways in which it is being damaged and then devise appropriate means of addressing them. In the first instance, this involves environmental management. Environmental management is a complex task, which incorporates many interrelating aspects that interface with environmental agencies, industrial units, corporate sector, consumers, governmental and non-governmental organisations etc. Every company needs an environmental management plan to help examine environmental risk and to ensure that a successful environmental management system is installed to minimise this risk.

Why Management of Environment?

Environment involves everything, living or non-living, that surrounds us, for example, air, land, water, people, plants, animals, minerals, automobiles, factories, etc., therefore, an overall system of management is required to integrate our efforts in this direction.

Everyone of us has a right to live in an environment of quality. Therefore, it is essential to understand the function and interaction of physical and biological elements of the environment and apply this knowledge in sound management programmes to conserve the natural resources and culture. Hence, management of renewable resources in industry, agriculture, recreation, health, forestry, fisheries, education or urban development is the primary need of the times. Although man has continuously tried to improve his understanding about his environment, his efforts to manage it have fallen short of preventing the deterioration and depletion of resources due to his continuous indulgence in the exploitation of nature for immediate gains. As such opportunities to design strategies for the effective management of our environment are unlimited. The complex environmental problems not only require imaginative and innovative specialists and technology but also the people's participation. Environmental management is "a process of planning, review, assessment, decision-making and the like, which is essential in the real-life situation of limited resources and changing priorities". This is not only essential for our present generation but also for future generations. The conservation of resources and maintenance of the quality of environment can be possible only through proper management. This is more true and essential for developing countries, where degradation of environment is going on at a faster rate due to overexploitation of natural resources and adoption of new technology.

The important aspects of environmental management are:

(i) to create a pollution-free environment;
(ii) to protect man and animals from pollution;
(iii) to protect the bio-diversity of the world;
(iv) to establish coordination between government and non-government organisations in protecting the environment;
(v) to analyse the impact of developmental plans on environment;
(vi) to help in formulation of national and regional environmental policies;
(vii) to monitor the ongoing plans for the quality of environment;
(viii) to sponsor awareness programmes;
(ix) to educate people at all levels, i.e., environmental education at different levels;
(x) to develop long-term and short-term plans for the conservation of the environment;
(xi) to examine the efforts made under environment management and its results and also fix responsibility for non-implementation, if any;
(xii) to encourage research in the various fields of environment; and
(xiii) to suggest guidelines to the government for the improvement of the quality of environment.

Approaches to Environmental Management

Environmental management has now become a priority area throughout the world because of the serious consequences that have either been experienced by mankind or are likely to face in future. There is a growing urgency of the management of land, water, minerals, soil, grasslands, forests, animal life, atmosphere, etc., or in other words, the whole environment. Any plan envisaging management of the environment must be based on the fact that each area has certain characteristics and can tolerate a range of physical conditions to a limited extent and every human activity affects the environment favourably or unfavourably. Therefore, two courses are open to us in using our surroundings:

(a) We adopt our needs and demands to the capabilities of each area for example, some recreations, urban development, waste disposal without overloading the water or the air.

(b) We modify an area to change its capabilities. To achieve this, we may take up *intensive management*, controlling the physical features of the environment (e.g., agriculture plantation, social forestry, pond-fish culture, etc.). In *less intensive* management, we modify the environment to favour selected types of living beings present in the area (e.g., wildlife management in sanctuaries, extensive forestry). Regulation of *harvest* from the available population (e.g., plucking or gathering fruits, firewood in the forest) is also possible.

The best approach to environmental management is an integrated approach in which all the components of the environment are taken into consideration and its proper management, as a whole, is done. While doing so, the following aspects should be taken into account:

1. Perception and awareness of environment:
   (a) source of perception and awareness,
   (b) level of perception, and
   (c) role of perception in environmental planning.
2. Environmental education and training:
   (a) at school, college and university levels,
   (b) through media, and
   (c) at research and training institutes.
3. Resource management:
   (a) classification of natural resources,
   (b) survey and evaluation of eco-resources,
   (c) conservation of energy, mineral, forest, soil and water resources, and
   (d) proper utilisation of human resources.
4. Environmental impact assessment:
   (a) assessment of present conditions,

   (b) assessment of the impact of industrial and technological developments,
   (c) assessment of the efforts done for eco-balance, and
   (d) development of eco-friendly techniques.
5. Control over environmental degradation and pollution:
   (a) purification of degraded environment,
   (b) pollution control,
   (c) monitoring, and
   (d) forecasting of natural hazards and to minimise the losses.

There are two approaches of managing the environment, viz., (i) preservative approach, and (ii) conservative approach. According to the first approach, man should not disturb the natural system and should adapt to it. But this is not possible because for all types of developmental activities he will have to use nature and its components, resulting in eco-imbalance of varied intensity. The conservative approach is that there should not be overexploitation of nature and conservation of natural resources is essential for sustainable development. In fact, proper utilisation and conservation of resources is the prime objective of environmental management. In the World Conservation Strategy (1980), the three main objectives stated were: the maintenance of essential ecological processes, the preservation of genetic diversity, and the sustainable utilisation of natural resources.

In recent years, several approaches have been developed for the management of environment. Regional planners often adopt a human ecology approach, while other planners suggest a system analysis or an ecosystem approach. Doxiadis (1977) has developed a science of planning settlement in balance with nature and termed it as *ekistics*. The various approaches developed for the environmental management are:

(i) *Ad hoc approach*: developed in reaction to a specific situation.
(ii) *Problem-solving approach*: for identification of problems and needs and implement solutions.
(iii) *Systems approach*: such as ecosystem, agro-ecosystem, etc. *Regional approach*: based on ecological zones such as watershed, river basin, coastal zone, command area development, island, etc.

(iv) *Specialist discipline approach*: often adopted by professionals for air, water and land management, urban management, tourism management, and environmental health.
(v) *Voluntary sector approach*: encouraged and supported by NGOs.
(vi) *Commercial approach*: for environmental management for business.
(vii) *Human ecology approach*: for study of relations between humans or society and nature.
(viii) *Political ecology approach*: to develop policies and laws.

Apart form the above, an environmental management system (EMS) approach has been developed as an integrated and proactive approach to environmental issues. It helps industry or other bodies and is designed to ensure that an environmental policy and environmental objectives are adopted and followed. Figure 4.1 illustrates a basic EMS approach developed by Hunt and Johnson (1995).

Figure 4.1
Environmental Management System

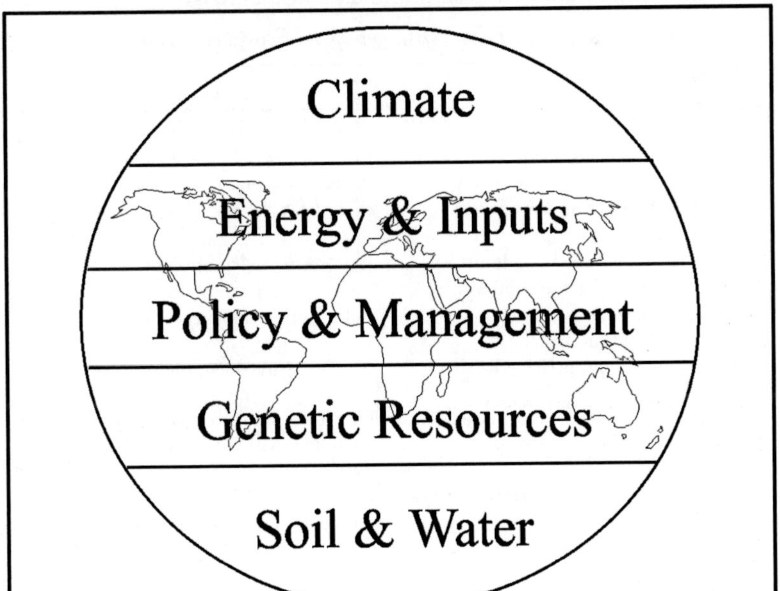

Thus, the EMS system approach helps to:
- develop a proactive environment approach;
- ensure a balanced view across all functions;
- enable effective, directed environmental goal setting; and
- make the environmental auditing process effective.

Ecosystem Approach

The ecosystem concept has become a widely used conceptual tool for research after 1945. Nowadays, ecologists often adopt an ecosystem approach when seeking to understand and monitor a given situation. The ecosystem concept allows the environmental managers to look at portions of complex nature as an integrated system. It may be applied to cities or agriculture (urban ecosystems and agro-ecosystems respectively). The ecosystem approach allows a holistic view of how components work together, in other words, it can incorporate human dimensions into biosphere functioning. It also helps define the temporal and spatial scale of management, and thus, is a multidisciplinary approach in order to deal with complexities of ecosystem function and usage. The ecosystem approach has both advantages and disadvantages as listed below:

Advantages and Disadvantages of the Ecosystem Approach

Advantages	*Disadvantages*
Comprehensive, holistic approach for understanding whole system.	May neglect socio-cultural issues such as politics, power and equality.
Draws on theory and methods from different fields to generate models and hypotheses.	Nebulous: a vague, superorganismic theory of poor empirical foundation.
Contributes to understanding limits, complexity, stresses and dynamics.	Non-standard definition of 'ecosystem'.
Encourages preventive thinking by placing people within nature.	Reflection of analytical systems.
Facilities locally appropriate, self-reliant, sustainable action.	Narrow spatial focus on local ecosystem.
Requires recognition of mutual dependence on all parts of system, e.g., natural/ cultural, person/family.	Duplicates and/or overlaps other disciplines without special contribution of its own.
Results in criteria for management actions.	If ecosystem approaches can apply to everything they may be meaningless.
Facilitates studies that integrate a range of disciplines.	

Based on D.S. Slocombe (1993).

Despite certain disadvantages, the ecosystem approach can best be used in (i) range management, (ii) forest management, and (iii) conservation management.

Human Ecology Approach

Human ecology is the study of relations between humans or society and nature, through a multidisciplinary approach. Similar to that of ecosystem approach, the main emphasis in this approach is on social relationship with environment which is a primary aspect of any management for planning and development. The scale of approach may be local to global, and it supports holistic study. The concept of 'Social Impact Assessment' (SIA) seeks to assess whether a proposed development alters quality of life and sense of well-being and how individuals, groups and communities will adopt to change caused by development. The socio-economic and biophysical aspects of the environment are interconnected, therefore, for environmental management, human ecology approach should be adopted along with political ecology approach.

Political Ecology Approach

Political ecology also studies relationships between society and nature. It holds that radical changes in human habits are required in order to counter environmental degradation and achieve sustainable development. These are likely to be different perceptions of environmental needs and problems between planners, policy makers, ministers, various departments of government, etc. All this can be effectively done by adopting political ecological approach of environment management.

Commercial Approach

Nowadays, there has been an increasing emphasis on environmental management for business as well as role of business houses in environmental protection. The emphasis is on:

- green corporate environmental management;
- green business ethics;
- eco-auditing;
- impact assessment, hazard and risk assessment;

- green marketing, labelling;
- recycling and waste disposal;
- environmentally sound investment and funding; and
- total quality management.

The commercial approach involves corporate priority, employee education, customer advice, transfer of technology, prior assessment, facilities and operations, research and compliance and reporting.

Integrated Environmental Management

In the present globalised world, the environmental problems have crossed political and geographical boundaries, therefore, there is need for integrated environmental management. The efforts should be properly coordinated between policy makers, planners, environmental scientists, businessman as well as public. The roots of integrated environmental management lie in integrated area development approaches and comprehensive regional planning. According to Born and Sonzogni (1995), the key elements of integrated environmental management are:

- coordinated control, direction, or influence of all human activities in a defined environmental system (such as a river basin or a watershed) to achieve and balance the broadest possible range of long-term and short-term objectives;
- a process of formulating and implementing a course of action involving human and natural resources in an ecosystem, taking into account the social, political, economic and institutional factors operating within the ecosystem in order to achieve social objectives; and
- an inclusive approach that takes into account the scope and scale of environmental and human issues and their interconnections.

Measures for Environmental Management

Environmental management is a complex system, which needs specialised knowledge, not only of the various facets of environment and associated problems but also the techniques of their application. Apart from trained personnel, environmental management is also related with proper perception of environment as well as with policy framework. Both governmental and non-governmental agencies have had their respective role in the management of environment. Similarly, public awareness is also essential for environmental management. The measures for environmental management vary from problem to problem. Apart from technical measures, there are certain general measures as under:

(i) Environmental education
(ii) Environmental legislation
(iii) Monitoring and mapping
(iv) Use of remote sensing
(v) Environment impact assessment
(vi) Environmental management plan

Environmental Education

One of the most important aspects of environmental management is environmental education. This mainly concerns itself with the various aspects of environment and ecology. Its primary objective is to teach the components of environment, ecology and ecosystems, man-environment relations, degradation of environment including various forms of pollution and their impact, environment vs. development, population, urbanisation and environment, environmental management and contemporary issues related to environment.

Although some aspects of environment were also a part of the several physical and social sciences, they do not provide an overall picture of this diversified subject. Therefore, during last 50 years, it has been realised that environmental education should be given not only to the students but to the general masses also, so that they can play a positive role in environmental protection. Environmental education is necessary because:

(i) it provides a complete knowledge of the components of environment;
(ii) the knowledge of ecosystems and interaction can be given;

(iii) it gives complete knowledge of various types of pollution, their causes, effects and measures of control;
(iv) it explains the role of man in degradation and maintenance of environment;
(v) it provides knowledge of the latest research in the field of environment;
(vi) it helps in environmental awareness;
(vii) it pursues the people's participation in environmental issues; and
(viii) it prepares research workers and social workers to work for the conservation of the environment.

In brief, environmental education is the need of the hour and priority should be given to it so that in phases the entire society will become aware of environmental issues. This will be the real success of environmental management programme. In 1981, the Indian Environmental Institute suggested an outline for environmental education. The main points were:

(i) The pattern of environmental education should be such that it could help in formulation of environmental policies and reflect the relationship between man and environment.
(ii) Environmental education should be included in secondary and university level syllabus, so that awareness among students can be raised.
(iii) It should also include various principles and concepts of environment.
(iv) At university level, knowledge of regional environment should be given and a report be prepared on any aspect of environment of that region by a group of students.
(v) Training to the students should be imparted so that they can educate the general public about environment.
(vi) An integrated rural development concept of environment should be applied.
(vii) Orientation and reorientation programmes should be organised not only for teachers but for doctors, engineers, planners, social workers and politicians also, to give them complete and up-to-date knowledge of environment.

Throughout the world environmental education has become very popular. There are several research institutions engaged in the field of research in environment. Similarly, several world organisations, national organisations and non-governmental bodies are engaged in environmental education. But still there is a long way to go in countries like India, where 72.2 per cent of the population is living in rural areas and their literacy level is very low. The media can also help in environmental education for the masses.

Environmental Legislation

The conservation and protection of environment in general and its specific variables like air, water, forests, sea, wildlife, etc., in particular is controlled through several international and national. laws. Thus, to implement these laws and develop public awareness about them is an important aspect of management. The laws to protect environment are essential, because with the growth of commercialisation and also due to man's greed, overexploitation of environment has become a common feature. This can be checked only through legislation.

Initially, the subject of protection of the environment was dealt with in a fragmentary manner under international law through various international conventions and treaties. In 1963, the Nuclear Weapons Tests Ban Treaty and in 1968 Treaty on the Non-Proliferation of Nuclear Weapons were signed. Treaties on Outer Space (1971), Oil Pollution in Seas (1969), etc., have also been made. A few important international conventions/laws are:

1. International Plant Protection Convention, 1951
2. The International Convention for the Prevention of the Pollution of the Sea by Oil, 1954
3. The African Convention on the Conservation of Natural Resources, 1968
4. The Convention of Wetlands of International Importance, 1971
5. The Convention of the Protection of the World Cultural and Natural Heritage, 1972

The year 1972 was a landmark in the field of environment, when United Nations Conference on the Human Environment was held at Stockholm (Sweden) from 5 to 16 June, in which Declaration on the Human Environment was adopted. This may be considered as the beginning of environmental movement in the world. UNO has also adopted a programme known as UNEP (United Nations Environmental Programme) and also "Earthwatch" and "Global Environment Monitoring System" and declared 5th June as "World Environment Day". A new international machinery known as "Habitat" was established in Nairobi (Kenya) and in 1982, the tenth anniversary of Stockholm Conference was celebrated in Nairobi, followed by the Nairobi Declaration. In fact, a number of conventions, treaties and instruments have been concluded or adopted dealing with the problem of environmental pollution in one way or other, which shows that all over the world environment and its protection has become an issue of concern and that all the international bodies have shown their concern about it.

Apart from international laws, every country has enacted laws regarding environment protection, pollution control, etc. In India, there are several Acts for environment protection. Even in Indian Constitution, three clauses, viz., 47, 48A and 51A, have been introduced for the protection of the environment. The four notable Acts of great importance are:

1. The Forest Conservation Act, 1980
2. The Prevention of Air and Water Pollution Act, 1974/1981
3. The Air Prevention and Control of Pollution Act, 1981
4. The Environmental Protection Act, 1986

It is the duty of the management personnel to have complete knowledge of various laws about environmental protection and help develop an awareness among the people about their provisions.

Monitoring and Mapping

Monitoring of environmental degradation and pollution and its associated aspects is essential for environmental management. Monitoring simply means checking of the environment, pollution, etc. In other words, quality of environment or quantity of pollutants is to be measured in order to find out the effectiveness of the measures taken for its improvement. Thus, monitoring is a continuous process and requires trained personnel. The five categories of monitoring are:

1. *Instrumental system* in which physical characteristics are measured by instruments, such as:
(i) sound level meter,
(ii) geiger counter,
(iii) thermometer,
(iv) oxygen meter,
(v) hydrometer, and
(vi) clorimeter.

2. *Chemical analysis* through mechanical devices is done by several methods such as:
(i) atomic absorption spectroscopy,
(ii) infra-red spectra photometry, and
(iii) gas-chromatography.

3. *Chemical analysis* in which volumetic, gravimetric and colourimetric analysis is done.

4. *Biological system* is done by examining the nature of certain plants and other micro-organisms.

5. *Hybrid system* is the combination of any two systems mentioned above.

Today, monitoring of environment has become popular even in developing countries because it helps in setting standards and limits of tolerance as well as for examining the effectiveness of the measures taken for improvement.

Use of Remote Sensing

One of the most exciting developments of modern times has been the advent of satellite remote sensing. Remote sensing simply means sensation received from a distance without direct contact. Technically, the term is used to denote the acquisition and measurement of data/information on some property(ies) of a phenomenon, object, or material by a recording devise not in physical, intimate contact with the features under surveillance. Thus, the informations, photos and imageries received through satellite remote sensing are very useful in environmental planning and management.

The practical use of remote sensing is mainly to oversee conditions or solve problems dealing with earth's environment and man-made infrastructures. The emphasis has therefore been on the status of land and water surfaces, and also on the overlying atmosphere, insofar as these affect human activities. The payoff from this monitoring is a better understanding of and interaction with (usually changing) events which influence our life, such as food production deforestation, urban expansion, mineral extraction, movement and severity of major storms, flooding, ocean nutrient detection, industrial site selection, occurring at both regional and local scales. Imageries received from satellite are very useful in disaster management as well as in management of land, forest, etc. Remote sensing, when combined with GIS (Geographical Information System), becomes a very useful system not only in studying the environmental problems but for their planning and management also.

Environmental Impact Assessment (EIA)

EIA is an important input of environmental management. It is a prior assessment of the future impact of the consequences of any decision on the quality of the total human environment on which man largely depends for his well-being. It is a systematic evaluation of all significant environmental consequences, and includes:

(i) prediction of the anticipated changes in the environment; and
(ii) determination of the magnitude or scale of the particular change, and application of an important significant factor for the change.

EIA is thus a systematic approach to environmental planning and management. A detailed analysis of EIA will be done in the next chapter.

Environmental Management Plan (EMP)

In the process of planning, it is essential for every project to formulate an Environmental Management Plan (EMP) to ensure that resources are used with maximum efficiency, and that each of the adverse impacts identified and evaluated as 'significant' be prevented, attenuated, or, when required, compensated. Possible mitigation measures include:

- changing project sites, routes, production technology, raw materials, disposal methods, engineering designs and safety requirements;
- introducing pollution controls, recycling and conservation of resources, waste treatment, monitoring, phased implementation, landscaping, inclusion of a plan for developing a green belt in an industrial plant site, personnel training, special social services or community awareness and education; and
- devising compensatory measures for restoration of damaged resources, monetary compensation for affected persons, offsite programmes to enhance some other aspects of the environment or quality of life for the community.

The identification, prediction and assessments are essential aspects of EMP. Identification of mitigation as well as enhancement measures based on speculation of the issues emerging out of the implementation of the project. EMP can thus include several technological and managerial interventions such as:

- recycling and conservation of resources;
- pollution control measures;
- phased implementation;
- monitoring;
- personnel training;
- landscaping (e.g., inclusion of a plan for developing a green belt around industry);
- devising compensatory measures for restoration of damaged resources;
- monetary compensations for affected persons; and
- off-site programmes to enhance some other aspects of the environment of quality of life for the community.

EMP for an industrial project should be based on identification of environmentally significant issues and recommendations of measures for each issue. It should be identified at the following stages:

- project siteing;
- site preparation and construction of the facility;
- operation of the plant; and
- project closure.

During project siteing, it should be ensured that alternative sites be selected being located in any of the following areas:

- ecologically sensitive habitats such as mangrove, estuaries, wetlands, coral reefs, etc.;
- watercourses, causing their eventual degradation;
- areas where meteorological and topographic conditions are conducive to temperature inversions and air pollution episodes;
- areas with significant environmental problems (air, water, and noise pollution); and
- areas with proximity to human settlements, resulting in impacts on human health.

Siteing should ideally be done in an area with proximity to raw materials, local workforce and transportation facilities. Sites selected should fulfil the following conditions:

- plot size should be sufficient for landfill of disposal on-site, or should be close to suitable disposal site(s);
- it should be convenient for public/private contractors of collect solid wastes for disposal;
- plot should be situated on a watercourse having maximum water dilution and absorbing capacity;
- it should be situated in an area where waste water can be reused, with minimal treatment, for agricultural or industrial purposes; and
- it should be situated within an area that is able to accept plant wastes in their sewage treatment system.

If possible, the project should be located in industrial zones where there is provision of adequate water supply, sewerage and waste water treatment facilities. Industries with gaseous emissions should be located at high elevations in an area not subject to temperature inversions, and where the prevailing winds are towards relatively unpopulated areas. Transport sector studies should be conducted and safe transport routes selected to reduce likely impacts from spillage. Contingency measures for spillages should also be drafted.

In preparation of EMP due consideration should be given to socio-economic issues. The EMP, if properly prepared and its suggestions implemented, will lead to the long-term planning for sustainable development.

References

Born, S.M. and Sonzogni, W.C., Integrated Environmental Management: Strengthening the Conceptualization, *Environmental Management*, 19 (2), pp. 167–181.

Hund D. and Johnson, C. (1995) *Environmental Magagement Systems*, McGraw Hill, London.

Slocombe, D.S., Environmental Planning Ecosystem Science and Ecosystem Approaches for Integrated Environment and Development, *Environmental Management*, 17 (3), pp. 289–303.

5

Environmental Impact Assessment and Eco-Auditing

The present world is facing multifarious problems of environmental degradation due to technological and industrial development as well as an explosive growth of population, which has caused enormous strain on environmental resources. Thus, there is an urgent need for maintaining a balance between the capacity of environment and the quantum of sustainable utilisation. It is possible only by understanding environment in its entirety and the principles of its scientific management. In the preceding chapter we have discussed various approaches to environmental management, and also a suitable approach for implementation. There is now a growing concern about 'Environmental Impact Assessment' (EIA) all over the world, specially after the U.S. National Environmental Policy Act of 1969 in which it was stated that all federal agencies should conduct an EIA on all major developments before approval would be given to developers. The present chapter has been designed to discuss the EIA and also major issues and strategies for proper environmental management.

Environmental Impact Assessment

Nature and Characteristics

EIA is the prior assessment of the future impact of the consequences of any decision on the quality of the total human environment on which man largely depends for his well-being. There is no universal definition of what exactly EIA is, but a simple explanation might be "an approach which seeks to improve development by a *priori* assessment". It is difficult to provide a clear definition of an EIA, although its characteristics can be listed as follows:

- It is a systematic evaluation of all significant environmental consequences.
- It is a structured, systematic and comprehensive approach.
- It is a process which forces developers to reconsider proposals.
- It is a process leading to a statement to guide decision-makers.
- It is a process which has the potential to increase developer's accountability to the public.
- It should be subject to independent, objective review of results.
- It should include clear statement of identified impacts as well as possible alternative development options.
- It is the study of a proposed development project on the environment.
- It considers both environmental and economic costs and benefits.
- It is a predictive mechanism and estimates the changes in the quality of the current environment as a result of the proposed action.

The growth of EIA is associated with the legislation passed in the United States on 1st January 1970, arising out of the National Environment Policy Act (NEPA) of 1969, requiring EIA for major projects. The complete report of the EIA is called an Environmental Impact Statement (EIS), but it is not necessary that all EIAs lead to an EIS. In general, an EIA is an objective evaluation of the environmental problems that may arise from the implementation of a project, and it should provide the public with this information, clearly and completely.

Structure and Stages

The structure of an EIA varies according to the project, location and the authorship. EIA, which deals only with project impacts, includes an accurrent environmental conditions (known as the baseline study) and accounts of project impact as well as available alternatives to the proposed project. EIA differs in both structure and their emphasis. A general design can be constructed by dividing EIA into several sections. These are as follows:

1. Preliminary activities
2. Scoping
3. Baseline study
4. Environmental impact evaluation
5. Mitigation measures
6. Assessment of alternative measures
7. Preparating of final document
8. Decision-making
9. Monitoring of project implementation and impacts

Figure 5.1

Represent the Stages Involved in EIA

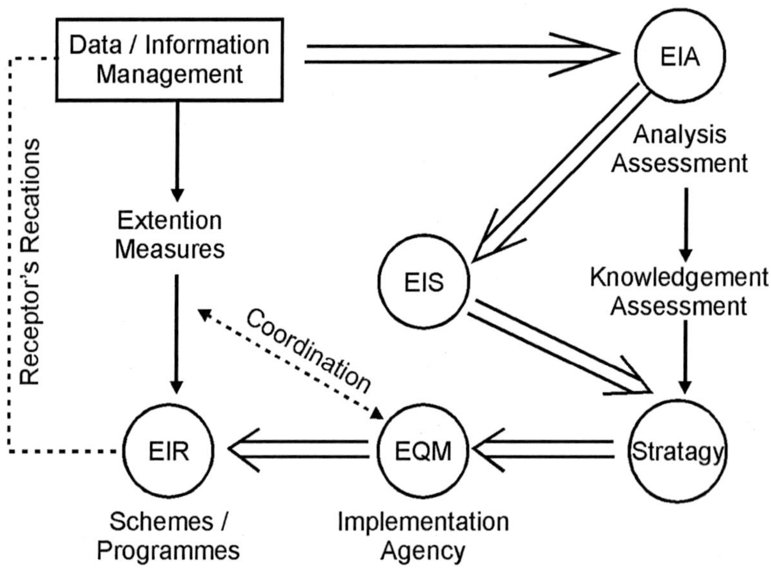

EIA = Environmental Impact Assessment
EIS = Environmental Impact Statement
EQM = Environmental Quality Management
EIR = Environmental Impact Remedies

A project can impact a given environment under three circumstances. In the first instance, it may arise from project location. This happens because the project would change the existing attributes of environment in the locality. Every such change would release specific process- bound changes. Secondly, some impacts may take place during construction of the project. These can produce temporary or permanent changes in the environment. Then, after completion of the given project, the manner of envisaged uses of it may cause impacts upon environment. All such expected impacts are to be stated in the EIA report. The essential steps in an EIA are:

(i) Prediction of the anticipated changes in the environment.
(ii) Determination of the magnitude or scale of the particular change.
(iii) Application of an important significant factor for the change.

Figure 5.1 represents the stages involved in an EIA.

Techniques

Various techniques are used for identifying environmental impacts. These include:

(i) checklists, (ii) matrix, (iii) networks, and (iv) overlay techniques.

Checklists used for identifying project impacts are expected to encompass all possible impacts of this kind. Table 5.1 illustrates a section of a checklist designed for a small water power project.

Earlier, simple checklists which only included a list of the environmental indices that could be affected were used. But, now detailed descriptions of the impact of each environmental factor are used in the form of questionnaire. In more sophisticated checklists, the relative importance of individual factors can be shown by varying weights being given to different factors along with the total ranking calculated for all the versions of the project.

Matrices are very suitable for EIA as they link a particular environmental factor to a specific action of the development project. Leopold and his associates, in the late 1960s, designed techniques for evaluating landscape. The Leopold matrix is a large arrangement with 88 environmental factors along the vertical axis

Table 5.1

Parts of a Checklist for Evaluating a Water Power Project

A. Physical Impacts

- change in sediment transport downstream
- change in water quality
- change in river hydrology
- river erosion
- slope instability
- deforestation
- increased erosion on slopes
- increased seismic disturbance in the area
- inundation of significant amounts of natural vegetation
- disturbance of the local fauna
- alteration of aquatic life
- alteration of soil fertility
- loss of rare or unique features or life forms

B. Impact on Local Population

- inundation of settlements
- resettlements of people
- problems of sanitation
- availability of river water downstream
- spread of endemic diseases
- introduction of new diseases
- psychological stresses due to changes in lifestyle and economy

C. Impact on Local Economy

- disturbances in local economy
- inundation of existing transport network disruptions in river transport
- submergence of objects of cultural heritage
- inundation of historicity sites
- availability of electricity
- improvement of agriculture
- industrial expansion

and 100 development characteristics along the horizontal one, giving rise to 8,800 cells. Each cell is divided by a diagonal line, and magnitude and importance of the impact are written on a scale 1 to 10. This type of matrix is called an interaction matrix. A part of Leopold matrix is given in Figure 5.2.

The Leopold matrix has following characteristics:

- It is excellent as a basic scanning tool.
- It allows the application of only a section of the matrix, as required.
- It can be used to indicate both beneficial as well as adverse impacts by writing a plus or minus sign in front of the entries in the cells.
- It can be used repeatedly throughout a project, to estimate the impact at different stages and also for different parts of the project areas.

There are also other types of matrix. At times, a particular development step may affect an environmental factor which in turn alters other factors. Such changes are examined by using a stepped interaction matrix.

Networks are used to show interrelationships among the different components of the environment of the area and also to indicate the flow of energy or impact throughout the environment such as an upland ecosystem or a drainage basin. Different types of network are known as sequence diagrams, directed diagrams or impact trees. Networks can be used to show both temporal and spatial flows of impacts.

Overlay techniques are used in planning before formal EIA process starts. It provide a base for the study of EIA. Individual impacts, such as affecting soil, water, settlements and noise conditions, are separately summarised and mapped over the area using choropleths to indicate the intensity of the impact. These maps are transferred on to transparencies which are then laid over one another to produce a composite effect, Summing up the total impact as shown in Figure 5.3 with the advent of Geographic Information System (GIS) and computer technology, this system has now become more useful.

Figure 5.2

Part of the Leopold Matrix, From Leopold, L.B. et al., (1971) A Procedure for Evaluating Environmental Impact, US Geological Survey.

1. Identify all actions (located across the top of the matrix) that are part of the proposed project.
2. Under each of the proposed actions, place a slash at the intersection with each item on the side of the matrix if an impact is possible.
3. Having completed the matrix, in the upper left-hand corner of each box with a slash, place a number from 1 to 10 which indicates the MAGNITUDE of the possible impact; 10 represents the greatest magnitude of impact and 1, the least, (no zeroes). Before each number place + if the impact would be beneficial. In the lower right-hand corner of the box place a number from 1 to 10 which indicates the IMPORTSNCE of the possible impact (e.g. regional vs. local); 10 represents the greatest importance and 1, the least (no zeroes).
4. The text which accompanies the matrix should be a discussion of the significant impacts, those columnns and rows with large numbers of boxes marked and individual boxes with the larger numbers.

	a	b	c	d	e
a		2/1			8/5
b		7/2	8/8	3/1	9/7

I. EXISTING CHARACTERISTICS AND CONDITIONS OF THE ENVIRONMENT / **II. PROPOSED ACTIONS WHICH MAY CAUSE ENVIRONMENTAL IMPACT**

A. PHYSICAL AND CHEMICAL CHARACTERISTICS	PROPOSED ACTIONS	A. MODIFICATION OF REGIME: a. Exotic flora or fauna introduction	b. Biological controls	c. Modification of habitat	d. Alteration of ground cover	e. Alteration of ground water hydrology	f. Alteration of drainage	g. River control and flow modification	h. Canalization	i. irrigation	j. Weather modification	k. Burning	l. Surface of paving	m. Noise and vibration	B. LAND TRANSFORMATION AND CONSTRUCTION: a. Urbanization	b. Industrial sites and buildings	c. Airports	d. Highways and brides	e. Roads and trails	f. Railroads	g. Cables and lifts	h. Transmission lines, pipelines and corridors	i. Barriers including fencing	j. Channel dredging and straightening	k. Channel revetment	l. Canals	m. Dams and impoundments	n. Piers, seawalls, Marinas and sea terminals	o. Offshore structures
1. EARTH	a. Mineral resources																												
	b. Construction material																												
	c. Soils																												
	d. Land form																												
	e. Force fields and background radiation																												
	g. Unique physical features																												
2. WATER	a. Surface																												
	b. Ocean																												
	c. Underground																												
	d. Quality																												
	e. Temperature																												
	f. Recharge																												
	g. Snow, ice and permafrost																												
3. ATMOSPHERE	a. Quality (gases, particulates)																												
	b. Climate (micro, macro)																												
	c. Temperature																												
4. PROCESSES	a. Floods																												
	b. Erosion																												
	c. Deposition (sedimentation, precipitation																												
	d. Solution																												
	e. Sorption (ion exchange, complexing																												
	f. Compaction and setting																												
	g. Stability (slides, slumps)																												
	h. Strees-strain (earthquake)																												

Figure 5.3
Use of the Overlay Technique in an EIA

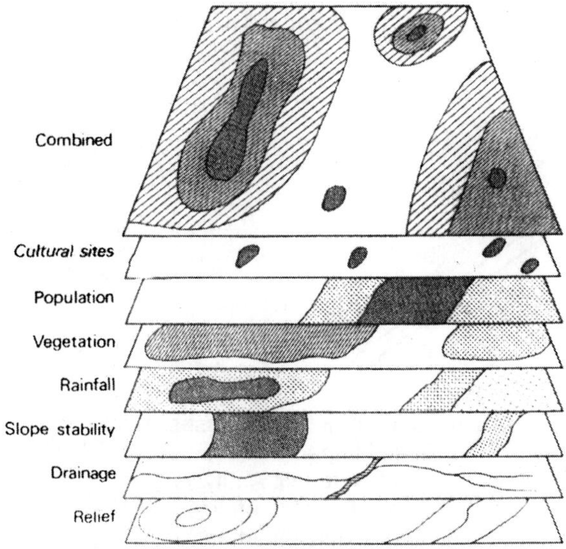

EIA and Planning

EIA is a systematic approach for planning. It can be a policy instrument and, as Graham Smith (1993: 12) argues, it should be treated as "a bridge that integrates the science of environmental analysis with the policies of resource management". Figure 5.4 below depicts how EIA fits into planning:

By improving understanding of the relationship between development and environment, EIA can lead to a better environmental management. If EIA is to become an integral part of planning, it must be applied before development decisions are made. While preparing the EIA report, it is useful to classify comprehensively the environmental resources upon which impacts are anticipated. These resources can be land (including soil), water (including surface and ground water), ambient air quality, ambient noise level, floral assemblages, faunal assemblages, human society and element of cultural heritage. For each category of resources to be impacted upon, the EIA report should specify the sources of the anticipated impacts and to suggest the required mitigation measures.

Figure 5.4
EIA and Planning

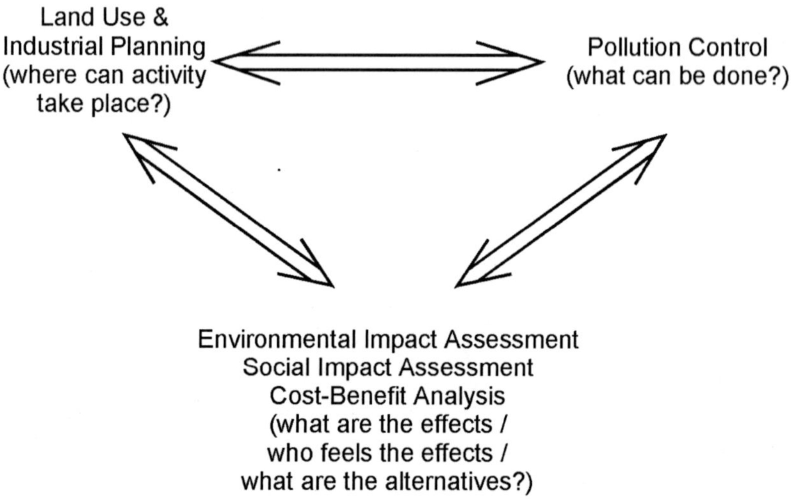

EIA cannot be taken in isolation, rather, it is related with other measures, i.e., state-of-environment reports, environmental audit, new system of national accounts, etc. This relationship has been shown in Figure 5.5.

The studies related to EIA are very useful in policy making, planning and administration. It should be related with Social Impact Assessment (SIA). The integrated regional environmental assessment is useful to:

(i) provide a broad, integrated perspective of a region about to undergo or undergoing developments;
(ii) identify cumulative impacts from multiple developments in the region;
(iii) help establish priorities for environmental protection;
(iv) assess policy options; and
(v) identify information gaps and research needs.

Figure 5.5
EIA and its Relationship with other Aspects

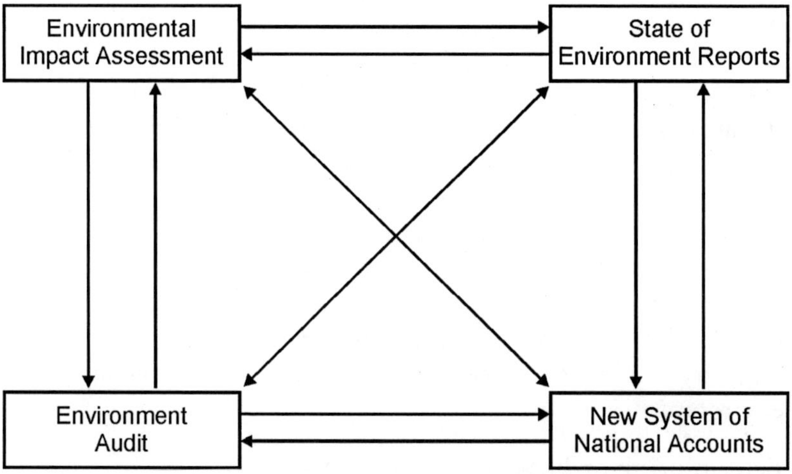

Problems

EIA is an essential component of a project but there are certain misunderstandings, which often elaborated by those who are not in favour of EIA. The three misunderstandings exist about EIA are: (i) EIAs have been interpreted as holding up development or at least delaying it. (ii) These are expensive exercises. (iii) This is just a bureaucratic process which does not really influence decision-making.

But none of these is correct. EIAs are designed not to delay development but to make sure that a large-scale development project does not degrade the environment severely and unnecessarily. EIAs are supposed to start very early at the pre-feasibility stage of the project, so that project design and EIA preparation can go simultaneously. Similarly, cost of EIA would range from 0.5 to 1 per cent of the project cost and this percentage comes down for bigger projects due to economics of scale. The cost of EIA is therefore reasonable, considering what it achieves in return. The third objection that it is simply a bureaucratic process is also not true, because EIA itself is not a formality but its recommendations are used in project design as well as by decision-making authorities.

The major problem with EIAs is size. Over time they have evolved into large reports which require the expenditure of considerable financial, technical and human resources. Such resources are not always available in a number of developing countries. In fact, all projects are not required intensive EIA. There are some donor agencies, such as the World Bank or UNDP, which provide financial assistance and some agencies have also designed concise environmental assessments, which are used to determine whether the project requires EIA or not.

Other practical problems that may come up include too many possible alternatives, a wide range of impacts, lack of necessary data, lack of technical expertise, difficulties with quantification of impacts, etc. But, in spite of these problems, EIA now has become a necessity for major projects, which have had their impact on environment.

Eco-Auditing or Environmental Audit

Eco-auditing or environmental audit is a process of extracting information about a company that provides a realistic assessment of how the company affects the environment and also a set of environmental objectives and targets to reduce the effects.

Eco-auditing is a systematic multidisciplinary method used periodically to assess the environmental performance of a project. Eco-auditing evolved as a management tool in the USA in 1980s. It has been promoted in Europe by the International Chamber of Commerce and by some multinational corporations as a means of getting effective environmental management. But, in developing countries, the eco-auditing concept is still a theoretical concept. However, India has modified its Companies Act to include a requirement for eco-audits.

The benefits are as under:

(i) it is a means for ensuring the continual improvement in environmental management;
(ii) it is a good method of monitoring;
(iii) it can assist efforts for sustainable development;
(iv) it can help involve the people in environmental management;
(v) it may reduce the need for government inspections;

(vi) it may help identify cost recovery through recycling and sale of by-products, etc.; and
(vii) it may generate valuable data for regional or national environment reports.

During the course of company's quest for an environmental system, two audits are done. The first is the environmental audit, which is done to know what a company actually does and how it affects the environment. The second audit is conducted to see whether an environmental system works according to its procedures and that objectives and targets are being achieved.

The eco-audit includes the following:

(i) internal audit;
(ii) external audit;
(iii) aspects and impacts analysis;
(iv) applicable legislations;
(v) environmental objectives and targets;
(vi) environmental policy statement ; and
(vii) corporate environmental plan.

The *internal audit* is the first step to list the key factors that the internal analysis might encompass and then conduct an internal appraisal. The results of this internal analysis are used to form the basis of aspects and impacts analysis. The factors and areas of the internal audit are:

(a) Financial:
—Liquidity and profit margins
(b) Personnel:
—Management experience
—Training and education
—Motivation and attitudes
—Workforce skills
(c) Research and development (R&D) and design
(d) Engineering and production
(e) Environmental management

The *external audit* encompasses all the forces and factors and factors outside a business that have some impact on it but that cannot influence or control it. In this audit, following points are taken into consideration:

(a) Environmental pressure groups
(b) Consumers:
- —Economic growth
- —Income levels
- —Interest rates
- —Exchange rates
- —Employment
- —Credit policies
- —Income distribution
- —Savings and debt
- —Taxation

(c) Competitors:
- —Age structure of population
- —Attitude towards health and lifestyle
- —Changing work pattern
- —Culture, etc.

(d) Operational policy:
- —Automation
- —New methods of travel
- —New materials
- —Improved communication

Aspects and impacts analysis is another phase of eco-auditing. ISO 14001 guidelines describe an environmental aspect as "... an element of an organisation's activities, products or services that can have a beneficial or adverse impact on the environment ..." similarly an environmental impact is "... the change that takes place in the environment as a result of the aspect ..."

This analysis is done with the help of well-prepared matrixes. The result of the analysis indicates:

- those areas of company's operations that may harm the environment;
- the different aspects of the company's operations have been classified into levels of environmental impact, i.e., low, medium or high;

— a scale 1–5 has been used to determine each impact's probability of occurrence; and
— a matrix has been created to calculate the impact scores of each aspect.

A *register of environmental legislations* is a fundamental part of the planning and implementation process. In this, all those regulations are mentioned that apply to the environmental aspects of the company activities.

The setting of *Environmental objectives and targets* is the main way in which a company's environmental performance can be improved. The environmental management plan is designed to be dynamic and open-ended list of objectives and goals for a company to achieve in accordance with its stated objectives and targets. The environmental plan should be regularly revived and updated by the environmental manager in consultation with the environmental working group.

Environmental policy statement states the company's commitment to the protection of the environment, prevention of pollution and continuous improvement of company environmental performance. For example, these aims could be to reduce:

— consumption of energy and water,
— generation of waste, and
— emissions.

The *corporate environmental plan* is the culmination of the auditing process. Having identified, listed, analysed and prioritised the operational aspects of the company, the environmental impacts, emergent targets and objectives will form the basis of the corporate environmental plan. Its prime objective should be the achievement of an environmental management standard.

The environmental audit will provide an insight into how a company operates. If an auditing process is already in place, it should be applied in environmental context. An audit is a natural management tool and it will become more effective when environmental audit is added to it. The need of the time is that all the developed and developing countries should adopt the eco-auditing system. For eco-audit, the establishment of standard is a prime task and most of the developed countries and a few developing countries have already established the 'standards' for the environmental protection, pollution management as well as for quality of life. In early 1992, the world's first eco-audit standard was published by British Standards Institute. A number of countries have adopted it which was revised in 1993 and 1994 to make it more compatible with the more recently introduced Eco-Management and Audit Scheme (EMAS). The International Standards Organisation (ISO) has also developed standards broadly compatible with EMAS. The need is that the system of eco-audit should be implemented as a part of environmental management.

References

Batisse, M. (1985), Action Plan for Biosphere Reserves, *Environmental Conservation*, 12(1), p.17.

Dimento, J.F. (1989), Can Social Science Explain Organizational Non-compliance with Environmental Law, *Social Issues*, 45(1), pp.109–112.

Goodland, R. (1985), *Wildland Management in Economic Development*, The World Bank, Washington DC, p.11.

Graham Smith, L.G. (1993), *Impact Assessment and Sustainable Management*, Longman, p. 13.

6

Standardisation of Environmental Management System

Environmental management is still a relatively young discipline, so judging how successful it has been and in what ways it should be 'tuned' to better serve the quest for development is difficult. Environmental management has to cope with natural threats and problems caused by human activities; it has to do this in a world where nature is being degraded, and it has to support livelihoods and steer these to ensure sustainable development. Environmental management deals with humans and natural processes, and it has to cope with changes in economy, technological capabilities, alterations of attitudes, social capital, social values, skills and many other variables. Therefore, for effective environmental management, there must be standardised management methods, techniques and tools so that the results can be easily and meaningfully compared with past studies or from elsewhere.

Some of the common tasks of environmental management and the selection of the tools used for the functions are as follows:

- Setting goals and objectives and strategic planning and assessment.
- Selecting methods and consulting guidelines and standards.
- Collecting relevant data, surveys, focus groups providing local knowledge, GIS, desk research, etc.
- Scenario development studies, environmental and social impact assessment, predictive modelling, Delphi technique, eco-foot printing etc.

- Providing support for environmental management and standardisation.
- Selecting enforcement methods, law, taxation, policing measures, incentives, education etc.
- Assessment of funding.
- Pilot studies, market research studies, etc.
- Public information and their reactions.
- Monitoring of ongoing measurement.
- Evaluation/auditing: eco-audit, sustainability audit.

The key principles of environmental management are prudence and stewardship, standards, monitoring and the other approaches for pursuing these principles.

Standards, Indicators and Benchmarks

A *standard* may be defined as a widely accepted or approved set of parametres of something against which others may be measured. Standards allow meaningful evaluation, exchange and comparison of data, improve objectivity of judgement, support negotiation, law making and comparison. Environmental standards may be divided into three groups as under:

— those concerned with ensuring human health and safety;
— those concerned with maintaining environmental quality; and
— those concerned with the quality of consumer items.

Standards and indicators are not static but change with time and technology. Standards have existed from ancient times. Even Egyptians, Greeks and Romans had units of measurement and medieval European crafts guilds set standards for quality of goods. By the nineteenth century, Britain, France and some other countries had set up institutes and observatories which developed, managed and regulated established standards. Standards are based on indicators, while benchmark is just a level that can be aimed for, or it can be a waymaker against which to judge standards or progress.

Setting of standards for environmental management is a difficult task, because environmental conditions have great spatio-temporal variations. As research into environmental issues progresses, new standards are needed. The process is ongoing, involving various natural and international institutes and standards organisations such as the British Standards Institution (BSI) or the International Standards Organisation (ISO). The standards in environmental issues also vary from country to country and with their level of development, but at the same time there are certain international standards which are necessary to follow so that a worldwide uniform pattern can be developed. Standards play a crucial part in:

- monitoring;
- modelling, to understand the environment and establish trends;
- negotiation;
- enforcement of rules;
- environment auditing; and
- maintaining environmental quality.

The fields of activity, which make use of standards, include:

- pollution control;
- the quest for sustainable development;
- health and safety;
- public hygiene and health (specially domestic water supplies, sewage and waste disposal);
- consumer goods (food standard, electrical safety, electromagnetic radiation safety);
- pharmaceutical products;
- transport safety and quality; and
- disclosure of information to the public.

On the other hand, *benchmarks* provide reference points by which to measure something, set minimum targets, and are a means for sharing and promoting food practices. Standards often rely upon *indicators* - things that can be easily measured, and which have specific meaning and point out something. Indicators are widely used to try and assess whether things are getting better or worse, including sustainable development. Some indicators are subjective and some objective. Ideally, an indicator should be sensitive, but not so much that it triggers false alarms; it should respond fast, reliably and unambiguously and if possible be cheap and easy to use. Indicators have been developed by most if not all fields: ecology, health care, pollution control, biodiversity assessment and conservation, social development, famine relief and many more. Indicators measure and warn of a huge diversity of risks, threats, quality changes, degree of sustainability, aesthetic value and so on.

Environmental Management System Standardisation

In 1990, the British Standards Institution (BSI) devised an environmental assessment standard based on the then existing quality standard BS 5750, now superseded by the ISO 9000 series of standards. After consulting industry, and following a 2-year pilot programme, in 1994, BSI launched the environmental standards BS 7750.

At the same time, the European Commission put forward a proposal for a scheme known as the Eco-Management and Audit Scheme (EMAS). The objective was to place the emphasis on the management of environmental systems and not just on the system itself. EMAS was launched in 1995.

The BS 7750 and EMAS standards were very similar in environmental requirements for businesses. At that time, it was thought appropriate to make EMAS a mandatory standard for businesses. Strong industry lobby groups argued successfully, however, that a mandatory approach would be detrimental to industry and EMAS is now a voluntary scheme. A brief introduction of both BS 7750 and EMAS is given below:

BS 7750

BS 7750 was the first UK national standard created for an environmental management system. Based on the BS 5750 quality system, the BS 7750 system was used to describe the company's environmental management system, evaluate its performance and define policy, practices, objectives and targets. This provides a catalyst for continuous improvement.

The concept is similar to that of BS 5750 and ISO 9000 for quality systems, in which the methods to be used are open to definition by the company. The standard provides the framework for development and assessment of the BS 7750 environmental management system. BS 7750 was developed as a response to concern about environmental risks and damage (both real and potential). Compliance with the standard is voluntary and complements the requirements for compliance with statutory legislation. As BS 5750 was the driver for ISO 9001, so BS 7750 led to the development of ISO 14001.

As its base, BS 7750 requires an environmental policy to exist within the company that is fully supported by senior management and that outlines the policies of the company not only to the staff but also to the public. The policy needs to clarify compliance with environmental legislation that may affect the company and stress a commitment to continuous improvement. Emphasis has been placed on policy as this provides the direction for the remainder of the management system.

The preparatory review and definition of the organisation's environmental effects is not part of a BS 7750 assessment, but examination of these data will provide an external auditor with a wealth of information on the methods adopted by the company. The preparatory review itself should be comprehensive in its consideration of input processes and output at the site. It should also be designed to identify all relevant environmental aspects that may arise from the company works. These may relate to current or future operations, as well as to the activities performed on site in the past, e.g., contamination of land.

EMAS

The Eco-Management and Audit Scheme (EMAS) is similar in structure to ISO 14001. There are, however, two major differences between the standards. The first is that the whole company can be certified to ISO 14001, whereas EMAS is generally a site-based registration system. The second is that, whereas any company from any business sector can use ISO 14001, EMAS is only available to those companies operating in the industrial sector. Within the UK, an extension to the EMAS scheme has been agreed for local government operations, which may also register their environmental management systems to the EMAS regulations.

In addition to a summary of the process, the statement requires quantifiable data on current emissions from the site and environmental effects, the amount and types of waste generated, raw materials utilised, energy and water resources consumed, and any other environmental aspects that may relate to operations of the site.

Pre-assessment is as much part of EMAS as it is of ISO 14001. The environmental audit must be comprehensive in consideration of input processes and output at the site. The procedure is designed to enable identification of all relevant environmental aspects that may arise from the site itself.

The pre-assessment will also include a wide-ranging consideration of the legislation that may affect the site, whether it is being compiled with currently, and perhaps even whether copies of the legislation are available. Many of the environmental assessments undertaken have already highlighted the fact that companies are unaware of all the environmental legislation that affects them, and, being unaware, they are often found not to be meeting the requirements of that legislation.

Under the EMAS standard, the company will declare its primary environmental objectives, i.e., those that can have most environmental impact. In order to gain most benefit, these will become the primary areas of consideration within both the improvement process and the company's environmental programme. The programme will be the plan used to achieve specific goals or targets along the route to a specific goal and will describe the real and achievable means to be used to reach those objectives.

As with ISO 14001, the EMAS standard requires a planned, comprehensive and periodic series of audits of the environmental management system to ensure that it is effective in operation, is meeting specified goals, and continues to perform in accordance with relevant regulations and standards.

In 1993, it was felt that an international standard was required for environmental management. Three years later, in 1996, from an idea based on BS 7750, the ISO 14001 standard was born. As BS 5750 had been withdrawn with the appearance of the ISO 9000 series, the outcome of the emergence of the international standard ISO 14001 meant that the national standard BS 7750—together with national standards in other EU countries—were, with common consent, also withdrawn.

Since the introduction of ISO 14001, many other ISO 14000 standards have also come into operation, as follows:

ISO 14001 Standards

Standard	*Description*
14000	Guide to environmental management principles, system and supporting techniques.
14001	Environmental management systems—specification with guidance for use.
14010	Guidelines for environmental auditing—general principles of environmental auditing.
14011	Guidelines for environmental auditing—audit procedures, auditing of environmental management systems.
14012	Guidelines for environmental auditing—qualification criteria for environmental auditors.
14013/15	Guidelines for environmental auditing—audit programmes, reviews and assessments.
14020/23	Environmental labelling.
14024	Environmental labelling—practitioner programmes—guiding principles, practices and certification procedures of multiple criteria programmes.
14031/32	Guidelines on environmental performance evaluation.
14040/43	Life cycle assessment—general principles and practices.
14050	Glossary.
14060	Guide for the inclusion of environmental aspects in product standards.

Acorn Project

The most recent project designed to assist SMEs to achieve ISO 14001 is a trial project called Acorn. This is a pilot project focused on introducing greater flexibility to those SMEs seeking to achieve the ISO 14001 standard. Launched in the summer of 2000, Acorn is a joint Department of Trade and Industry (DTI) and BSI pilot study to test a five-step model for SMEs to achieve ISO 14001.

The five steps are as follows:

Step 1	A commitment to the ISO 14001 standard.
Step 2	Compliance with legal and regulatory requirements.
Step 3	Identification of significant environmental aspects and impacts.
Step 4	Management of significant environmental aspects.
Step 5	Documentation and integration of environmental management system.

As major companies are coming under great pressures to ensure that their suppliers also have the ISO 14001 standard, the Acorn project is designed to assist SMEs in meeting major company environmental demands. SMEs do not necessarily have to achieve full certification before becoming part of a major supply chain. Stage 3 of the project is the first validation and audit point for the standard; for most major companies, those SMEs achieving this stage will be accepted as a supplier.

ISO 14001 Self-Assessment

Before undertaking to adopt ISO 14001 environmental standard, it is necessary to know the present status as well as a fair understanding of the environmental terminology through a self-assessment questionnaire. Whichever assessment company you choose to achieve the environmental standard, the pre-assessment questionnaire should follow a similar format. It will comprise a number of sections relating to your company and how it interacts with the environment. The following list indicates the areas to be assessed:

- the company
- air emissions
- water pollution
- solid and hazardous waste
- soil and groundwater protection
- noise control
- resource management
- management systems
- other environmental impacts
- environmental exercise

The Company

This section requires you to supply general information about the company, e.g., name, address, number of employees and the nature of the business. State also the company's main business activity. If the company has more than one site, provide the same details for each site. If your company is aiming for the ISI 14001 standard, all company sites must comply with the standard requirements.

Air Emissions

This element asks you to consider which of your company's processes or business activities produce emissions to air. Consider everything vented, from heating fumes to incinerator flue gases, and record it. It is not necessary to be fully aware at this stage of the emissions being vented; this can be set down later by a third party if your company does not possess the expertise in house. You should be aware that, within the pre-assessment questionnaire, your company will be asked to provide answers to simple questions such as, "do you monitor air emissions?", "who is responsible for air emissions?", "are they audited?" and "is there a site plan showing emission points?" These are general 'yes' or 'no' questions that will provide the company and its assessor with the basic information needed to consider whether or not air emissions are to be an issue when going for full certification (Figure 6.1).

Figure 6.1
Air Emissions

Type of emission	*Activity*	*Site legislation/regulation/consent*

Who is responsible for managing air emissions?		
	Yes	No
Do you monitor air emissions?	❑	❑
Do you have objectives and targets for air emissions?	❑	❑
Do you have documented procedures for identifying, monitoring and controlling air emissions?	❑	❑
Have you conducted training for air emission procedures?	❑	❑
Have you audited these procedures?	❑	❑
Do you have a site plan illustrating emission points?	❑	❑
If so, where is this plan kept?	❑	❑

Water Pollution

A business is likely to use a significant amount of water, in everything from toilets to cooling processes. Make a log of everything that uses water. Make special note of those processes or activities that discharge water to the foul sewer, soakaways, other drains or natural streams, etc. With the same tick box 'yes'/'no' response, questions will be asked about wastewater objectives and targets, e.g., "do you have documented procedures and is there a drainage plan of all water usage and flows?"

Figure 6.2

Water Pollution

Type of discharge	*Activity*	*Method of discharge*	*Site legislation/regulation/ authorisation/consent*

Who is responsible for managing your wastewater discharge?

	Yes	No
Do you monitor wastewater discharges?	❑	❑
Do you have objectives and targets for wastewater discharges?	❑	❑
Do you have documented procedures for identifying, monitoring and controlling wastewater discharges?	❑	❑
Have you conducted training for wastewater discharge procedures?	❑	❑
Have you audited these procedures?	❑	❑
Is any wastewater discharged directly into controlled waters?	❑	❑
Is stormwater on your site discharged directly into controlled waters?	❑	❑
Is any wastewater treated on you site?	❑	❑
Do you have a site plan illustrating draining plans and discharge points?	❑	❑

If so, where is this plan kept?

Solid and Hazardous Waste

Your company also has a requirement to identify any waste that it produces. There are usually four main areas:

Figure 6.3
Solid and Hazardous Waste

Do you produce or handle any of the following?	Yes	No
Solid waste	❑	❑
Hazardous waste	❑	❑
Special waste	❑	❑
Clinical waste	❑	❑

Please complete if you produce any solid or hazardous waste:

Waste type	*Activity creating this waste*	*Method of disposal*	*Site legislation/ regulation/licence*

	Yes	No
Do you monitor your waste streams?	❑	❑
Do you have objectives and targets for dealing with waste?	❑	❑
Do you have documented procedures for identifying, monitoring and controlling waste?	❑	❑
Have you conducted training for your waste management procedures?	❑	❑
Have you audited these procedures?	❑	❑
Do you hold your own waste management licence?	❑	❑
Do you hold copies of licences for your waste transporters or disposal sites?	❑	❑
Do you have a region of consignment/transfer notes?	❑	❑
Do you have a site plan of waste generation and storage points?	❑	❑

1. solid waste—e.g., paper, plastic, metal, wood, etc.
2. hazardous waste—e.g., sludge from paper manufacture, paint-spraying residues, etc.
3. special waste—this has a lengthy legal definition that you are directed to if you are unsure of your company's waste classification, and will include chemicals and other substances that are highly flammable, irritant, harmful, toxic, carcinogenic or corrosive.
4. clinical waste—again, you will need to reference a legal definition if your company produces clinical waste; put simply, this covers medical waste, including instruments and diseased waste.

Soil and Groundwater Protection

If your company stores chemicals, solvents, paints, diesel, petrol, heating oils, etc., be aware of the consequences of any spillage. The emphasis here is on what the company stores and on where, and how, it is stored. If spillage of hazardous materials occurs, you will need to know what has been spilt, the volume of the spillage, where the spillage will drain to or flow naturally, and where it is likely to end up. A plan of the storage areas and directions of flows in the event of any spillage will be required. Any drainage plan will again be of assistance here.

If some water or land contamination does arise, you will need to identify the method you intend to use for remediation and methods for future protection. If your company currently uses waste contractors, they also need to be identified.

Noise Control

Check noise levels internally and externally. Most modern internally operated machinery generates excessive noise, even if it is only intermittent. External noise is also classified as pollution, and you can be sure that if your company is emitting excessive external noise, your neighbours, whoever they may be, will not be slow in telling you and the local authority and Health and Safety Executive (HSE). List those processes or other business activities that emit noise and have them measured by a third party to check the limits.

A check on the acceptance threshold levels can be actioned through your company's local environmental health department.

Resource Management

This is primarily a look at other resource-saving activities that may have been tested or implemented, e.g., to better manage fuel, energy and water consumption within the company. Take into consideration any waste or energy minimisation surveys that may have been undertaken directly by the company or indirectly by a consultant or other third party. You will be asked whether your company has waste or energy minimisation objectives and targets, whether it has a responsible person monitoring energy use and whether it has any material recycling or reuse initiatives currently in operation. You will also be asked whether your company has considered other uses for the land it occupies.

You may surprise yourself in this section by ticking more 'yes' than 'no' boxes. Most companies tend to focus on various methods of corporate cost savings and most would acquit themselves better under this section than in others.

Other Environmental Impacts

If you feel that previous sections have not really covered every activity that your company is involved in, this section allows you to consider the environmental impact that things such as odour, dust, vibration and visual impact may have on the environment.

Apart from the direct environmental impacts that your business activities may have on the environment, you will be asked to consider also the indirect impact. The indirect impact may be difficult to spot initially but can come in the form of, for example, product designs. Better product design may result in energy savings or a greater percentage of the product being recyclable. Better design of some brands of motor car has enabled a greater proportion of the vehicle—up to 85 per cent in some cases—to be recycled or reused.

Another area to consider is the packaging of distributed goods. You will be asked whether your company reuses the packing materials saved from incoming goods and whether your company uses its own transport for delivery. The use of a courier may be a worthwhile option with regard to reducing air emissions and fuel bills. One of the ways towards resolving a proportion of these indirect impacts is to ask your company's suppliers and customers whether there are more environment friendly alternatives to existing practices.

Charter for Sustainable Development

The ultimate goal for corporate environmental management systems should be to achieve a situation where a company's activities are in balance between continuous economic growth and caring for the environment. The International Chamber of Commerce (ICC) has developed a business charter for sustainable development. This effort has the support of many companies and is intended to identify areas for successful and continuous improvement. The charter promotes environmental policy areas as follows:

Corporate Priority

The aim here is for environmental management to become one of the highest priorities for all companies. This should be demonstrated by the creation of policies, programmes, procedures and practices leading to environmentally sound operations for the company.

Integrated Management

Full integration of environmental elements into the normal business operation of a company is essential to ensure that the management system functions as a coherent entity. The strength of this integrated approach can be demonstrated further by amalgamation of the management systems for environment, quality, customer satisfaction, and health and safety of staff. In effect, environmental management cannot be achieved by one person in a small office at the rear of the company.

Process of Improvement

The process of continuous improvement, from a sustainable development viewpoint, requires more than the operations of a company to be environment-friendly. It requires consideration to be given to how the company responds to customer expectations, community expectations, legal regulations, development of tools and new techniques in order to improve continually the environmental performance of the organisation.

Staff Education and Training

The development of a culture of continuous environmental education and training will provide greater awareness capability and motivation to employees and enable them to conduct their daily operational activities in an environmentally responsible manner. Such a programme can be a strong motivational tool and can bind employees closer to the company's environmental ethics.

Products and Services

The achievement of sustainability requires consideration of the life cycle of products and services (the term 'cradle to grave' is used extensively) to ensure that they have no undue environmental impact and are fit for their intended use, are efficient in their consumption of resources, and can be recycled, reused and disposed of safely.

Facilities and Operations

Similar to the 'cradle to grave' concept for products and services, facilities and premises can be renewed, regenerated or upgraded. When looking at existing plant and facilities, a stronger emphasis can be placed on the promotion of renewable resources, minimisation of pollution, the generation of waste, and safe and responsible husbandry of the land resource.

Contractors and Suppliers

The company should encourage contractors and suppliers to adopt sound environmental principles combined with a system of continuous improvement in line with or supported by its own system. Greening the supply chain ensures that the suppliers' objectives become consistent with the aims of the company itself. This also encourages the wider adoption of environmental improvement practices among suppliers and sub-contractors. The bigger the company, the easier and more readily this concept will be accepted by suppliers and sub-contractors.

Openness to Concerns

Sustainable development requires a company to become more transparent in all its business operations. This open approach is designed to foster an open exchange of information and communication between the company and its employees, suppliers, stakeholders, etc.

Compliance and Reporting

As part of a drive to be more transparent, a company pursuing sustainability will be required to measure and publicise environmental performance information. Following the publication of a company's environmental aspects and impacts, measurements can be made to demonstrate stability or improvement, and to confirm compliance with company objectives and legal requirements. This information is produced annually to provide information to all stakeholders.

In 1992, *Agenda 21* called for the establishment of indicators for sustainable development. Sustainability indicators, if they highlight the real underlying causes of environmental damage, will help prevent wasted efforts treating symptoms. A number of measuring indexs have been developed for sustainable development, such as Index of Sustainable Economic Welfare, Net Primary Productivity, Environmental Space, Factor X Concept, Composite Index of Intensity of Environmental Exploitation, Indicators of Farm Level Sustainability, An Extension of the Human Development Index, eco-footprint, etc. Environmental auditing, environmental accounting and eco-auditing as well as surveillance will help in developing an efficient environmental management system.

7

Air Pollution and Its Management

Air, which is a mixture of gases, moisture and some inert material, controls life on earth. It is a reservoir of oxygen needed by man and other animals and of carbon dioxide essential for plants. Any contamination in the air may disturb the whole atmospheric system, which is an insulating blanket around the earth. Without air, there would be no clouds, no winds, no rain, no snow and no fire. In other words, there would be no life on earth. The atmosphere, which shields the earth from ultraviolet radiation, becomes progressively tenuous with increasing distance from the earth. About 75 per cent of earth's atmosphere lies within 16 kilometres from the surface, and 99 per cent of it lies below an altitude of 30 kilometres.

The structure of the atmosphere is controlled by various gases present in the air. The clear, dry air at sea level contains 78 per cent nitrogen, 21 per cent oxygen, 1 per cent argon, and 0.03 per cent carbon dioxide and other gases (see Table 7.1).

All the major as well as minor components of the air/atmosphere are biologically important and the participation of each in the living processes is in some cases critically sensitive to slight changes in concentration. Whenever the proportion of the components is disturbed by man, it becomes a cause of air pollution, which nowadays is a major global problem.

Table 7.1
Composition of Clear, Dry Air at Sea Level

Components	*Per cent by Volume*
Nitrogen	78.084
Oxygen	20.9476
Argon	0.934
Carbon dioxide	0.0314
Neon	0.001818
Helium	0.000524
Methane	0.0002
Krypton	0.000114
Sulfur dioxide	0 to 0.0001
Hydrogen	0.00005
Nitrous oxide	0.00005
Xenon	0.0000087
Ozone	0.000007 (Summer)
	0.000002 (Winter)
Nitrogen dioxide	0 to 0.000002
Iodine	0 to 0.000001
Ammonia	0 to trace
Carbon monoxide	0.1 to 0.2 ppm

Source: H.J. Sanders(ed.), *Chemistry and the Atmosphere*, The American Chemical Society.

Nature of Air Pollution

In an ecosystem, all the elements act and react in an integrated form; therefore, this life supporting system continues in a systematic way. But, when any disturbance is created by man even in a micro form, there is a case of pollution. Growing industrialisation and transportation and the increasing use of pesticides and unwanted chemicals in the air have rendered the whole atmosphere polluted and its impact is very dangerous not only on man and other living organisms but also on environment itself.

Air pollution may be defined as imbalance in the quality of air so as to cause ill-effects. According to Maxwell (1973), "our enormously accelerated abuse of the atmosphere has become a health hazard and a threat to life, damaging both plants and animals in areas polluted with poisonous fumes, dust, and smoke." The industrial, automotive and domestic activities have resulted in increasingly outrageous insults to the atmosphere. Strahler and Straher (1977) write: "In air pollution the pollutants include gases and solids and liquid particles of both .organic and inorganic chemical classification." The different types of pollutants are continuously introduced into the atmosphere and are removed by natural processes of cleansing. But, when pollution exceeds the atmosphere's self-purifying capacity, accumulation of pollutants occurs causing serious hazards for environment and other organisms, including humans. The following are the main reasons for the increasing rate of air pollution:

(i) Poisonous gases and other particles emitted from industries without any treatment.
(ii) Heavy increase in the number of automobiles and their emission (smoke and other dangerous gases).
(iii) Increased use of chemicals and petrochemicals.
(iv) Population concentration in cities.
(v) Fast rate of deforestation.
(vi) Tests of experiments of atomic weapons.
(vii) Tests of chemicals and biochemical weapons.
(viii) Space research and satellite wastes.
(ix) Unorganised mining and traditional practices of the use of fuelwood, etc.

Certain natural events, such as volcanic eruption, dust storms, etc., are also responsible for air pollution. In order to understand the nature of air pollution, it is necessary to know the various sources of air pollution.

Sources of Air Pollution

Air pollution is the result of the combined effects of several pollutants. These pollutants are associated with each other and also react with other elements, therefore, it is difficult to categorise them, but for proper understanding they can be divided into the following categories on the basis of their origin, nature, size, impact, etc.

1. According to origin, particulate matter can be divided into two types, viz., natural and man-made. The natural form of particulate matter is the result of volcanic dust and gases, mineral dust and sea-salt crystals. Another class of natural particulate matter is smoke from forest and grass fires. Living plants release pollens and spores into the air, which are organic compounds. From forest trees certain hydrocarbons called terpenes are also released into the atmosphere.

Man-made particulate matter comes from many sources, but its major source is the combustion of hydrocarbon fuels—petroleum products, coal, peat and wood. Combustion of solid wastes is another source. Other kinds of matter introduced into the atmosphere are industrial chemicals, fly ash, refining fossil fuels, mining and smelting ores, as well as pollutants discharged through quarrying, farming activities, etc. Use of various types of solvents and also radioactive elements are the cause of air pollution created by man.

2. Another classification, according to origin, is (i) primary, and (ii) secondary pollutants. The primary pollutants are those gaseous and other solid micro particles inducted to the atmosphere. These pollutants are emitted and as such are not found in the air. The most important gaseous pollutants are carbon monoxide, oxides of sulfur, hydrogen sulfide, hydrocarbons, oxides of nitrogen, ozone and other oxidants. Secondary pollutants are the result of chemical reactions. Evidences and experiments indicate that exhaust gases of automobiles contribute more in the formation of secondary pollutants. For example, oxides of nitrogen produced in the combustion of petroleum and other fuels emitted into the atmosphere, yield ozone in the presence of sunlight. It is to be noted that ozone is not emitted as such into the atmosphere but formed only from primary pollutants.

3. According to chemical composition, air pollutants can be divided into organic and inorganic pollutants. Others have divided them into solid, liquid and gaseous pollutants. The gaseous pollutants are carbon monoxide, sulfur dioxide, hydrogen sulfide and organic sulfide, hydrogen fluoride, hydrogen chloride, oxides of nitrogen, aldehydes and organic acids, etc. Particulate pollutants consist of both solid and liquid particles. They vary in size from

0.01 micron to 20 microns. Dust, fume, mist, spray and smoke are included in this category.

4. According to source type, pollutants can be classified as being produced from:

(i) combustion,
(ii) transportation emissions,
(iii) industrial processes,
(iv) use of solvents, and
(v) radioactivity.

Combustion processes yield particulates such as fly ash and smoke and oxides of sulfur and nitrogen. The amount of sulfur dioxide emitted depends upon the contents in the fuel. High temperature processes such as thermal fixation of atmospheric nitrogen yield larger quantities of oxides of nitrogen. Carbon monoxide is also emitted from combustion. The other contaminants that are produced by combustion include acids and aldehydes. The simplest form of combustion is the use of fuels in domestic use. In most developing countries, including India, wood, coal, cow-dung and kerosene are commonly used for cooking. All these materials yield carbon dioxide, carbon monoxide, sulfur dioxide, etc.

Automobiles may be considered as the main source of air pollution, specially in urban areas. Automobile exhausts release smoke and to a certain extent lead particles also. Smoke contains the gaseous pollutants carbon monoxide, hydrocarbons and oxides of nitrogen. The pollutants emitted by petrol and diesel driven vehicles are as follows:

Table 7.2

Pollutants Emitted by Petrol and Diesel Driven Vehicles

Emissions	*Petrol Engine*	*Diesel Engine*
Carbon monoxide	3	—
Sulfur dioxide	40 ppm	400 ppm
Nitrogen oxides	1100 ppm	300 ppm
Unburnt hydrocarbons	150 ppm	20 ppm
Particulate matter	0.1 g/m^3	0.01 g/m^3

The rapid rate of industrialisation has resulted in more and more air pollution. Various industrial processes release almost all types of pollutants into the air. Some industries like cement, iron and steel, fertiliser, petrochemical, etc., are of great concern because of the difficulty in controlling the emission of pollutants from them. Acid rain has become a great threat to the environment. The use of solvents is increasing with the growing use of paints, spray, polish, etc. Due to the presence of hydrocarbons in these materials, air pollution is caused, which is dangerous for health. Similarly, spray of pesticides in agriculture is also responsible for air pollution even in rural areas.

Nuclear material, when released into the air, is hazardous for all living organisms. Nuclear weapon testings, nuclear reactors, chemical processing plants, research institutes and hospitals contribute many radionuclides to the atmosphere.

Specific Phenomena Related with Air Pollution

Ozone

Ozone is normally present in the atmosphere at about 0.05 ppm at sea level. It is produced naturally in the atmosphere by the action of electric discharges on oxygen. Ultraviolet radiation in the wave length of 260 mm is absorbed by oxygen molecules, causing them to react in this manner. Radiation from the sun probably causes high concentrations of ozone in the outer portions of the earth's atmosphere. Ozone plays a key role in the photo-chemical formation of atmospheric pollutants as well as being end product in the intricate complex of reactions that take place. The ozone layer provides a cover to the earth's surface and protects the earth from ultraviolet rays.

Until 1974, atmospheric scientists were proceeding in their research on possible impacts of nitrogen oxide (jet engine exhausts) on the ozone layer. But later it was found that a new threat to the ozone layer has emerged from synthetic chemicals called chlorofluorocarbon. This is a simple compound of chlorine, fluorine and carbon. The chlorine can destroy ozone. Recent studies indicate that if the production of chlorofluorocarbon continues at the present rate, the compound will enter the stratosphere in quantities capable of seriously depleting the ozone layer, normally termed as 'holes in ozone layer'.

The use of chlorofluorocarbon is increasing because of the demand of 'personal care products' such as deodorants, hair sprays, shaving creams and countless other consumer cosmetic products as well as in refrigeration. The ozone layer serves as a shield, protecting the troposphere and earth's surface from most of the ultraviolet radiation found in the sun's rays. If these ultraviolet rays reach the earth's surface in full intensity, not only all exposed bacteria would be destroyed, but plants and animal tissues would also be severely damaged. In this protective role, the presence of the ozone layer is essential in man's environment.

Greenhouse Effect

The temperature at the surface of the earth is maintained by the energy balance between the sun's rays that strike the planet and the heat that is radiated back into space. Some of the heat is absorbed and retained by the earth or objects on the surface. Much of this does not pass through the air envelope to outer space but is absorbed by the carbon dioxide and water vapours in the atmosphere and adds to the heat that is already present. Thus, carbon dioxide acts like the glass of a greenhouse, and on a global scale, tends to warm the air in the low levels of the atmosphere. This is called the greenhouse effect, which is also responsible for the increase in temperature over the earth's surface.

Scientists are of the opinion that carbon dioxide will increase global temperature significantly. There is no contrasting opinion about the fact that due to industrial and other forms of air pollution the carbon dioxide content is increasing. Volcanic eruptions are also responsible for the increase in carbon dioxide. Volcanic activity during historical period as well as in present times has thrown out dust and ash which spreads throughout the entire atmosphere and lasts for years.

There is evidence that the temperature of the entire earth has risen slightly during recent decades. Between 1885 and 1940, global warming was a worldwide phenomenon. There was an increase in mean annual air temperature of about 0.5°C. The greatest increase was in the northern hemisphere between 40°N and 70°N lat, where it was an increase of 1.6°C in the average winter temperature. After 1940 this process of global warming had become much slower, but with the increase in carbon dioxide content in the atmosphere from combustion, this has once again on the move. This will result in the receding of glaciers and melting of snow and raise the sea level. That may destroy not only the coastal regions but also other parts of the earth.

With the increase in temperature, scientists have predicted that the process of melting of ice caps of Antarctic, Greenland, etc., may lead to a rise in the sea level. Some of them have pointed out that there can be a rise of four feet in every ten years. There is a crucial relationship between the presence of the polar ice cap and global temperatures and if it happens its impact on climatic conditions of the world will be drastic and will change the entire pattern of our ecosystem.

Acid Rain

Acid rain is the outcome of sulfur dioxide (SO_2) gas, released into the air by the combustion of fossil fuels. This readily forms sulfuric acid (H_2SO_4) in the minute water films of suspended droplets in the air over cities. The washout of sulfuric acid by precipitation results in rain water with an abnormally high content of the sulfate ion, a condition known as acid rain. Nitric acid, formed by reactions involving the pollutant nitrogen oxides, may also contribute to the acidity of rain water.

The degree of acidity is measured in terms of pH value of rain water. The pH of rain water, as measured in the laboratory, is 5.65 owing to the dissolution of atmospheric CO_2 in it. Rain, having a pH value of less than 5.65, the value resulting from the equilibrium between atmospheric CO_2 and pure distilled rain water at 25°C, can be termed acid rain.

In recent years, it has been pointed out by water chemists that pH value of rain in north-western Europe and USA is going down to 3 and 5, which is an indication of acidity in rain water. The pH numbers are on a logarithmic scale, these values mean that rain in these areas has now often 100 to 1,000 times more acid than before. After studying rain water for 20 years the US scientists have reported average pH value of about 4. They have also observed values as low as 2.1 in rain water at certain localities. The acid rain is now not only the problem of industrial developed world but is also seen in countries like India. The observations of Khemani and his associates (1989) have been summarised in Table 7.3.

Some of the adverse environmental effects of acid rain may include excessive leaching of nutrients from plants, foliage and the soil, disturbance of balance of predators and prey in aquatic ecosystems, various metabolic disturbances to organisms, acidification of lakes and streams and the corrosion of structures.

Table 7.3

Average pH Values of Rain Water over Cities in India

High pH Values (> 7.0)	*Medium pH Values (7.0–6.0)*	*Low pH Values (6.0–4.5)*	*Very low pH Values (< 4.5)*
Ahmedabad	Agra	Trivandrum	Chembur(Bombay)
Allahabad	Bhopal	—	—
Amritsar	Calcutta	—	—
Bikaner	Chandigarh	—	—
Jaisalmer	Delhi	—	—
Jodhpur	Nagpur	—	—
Lucknow	Visakhapatnam	—	—
Srinagar	—	—	—

Source: Khemani, L.T. et al. (1989) Spread of Acid Rain over India, *Atmospheric Environment*, Vol. 23, pp. 757–762.

Smog

'Smog' is the name given for 'smoke fog'. The term was first used by H.A. Des Voeux in 1911 in a report on smoke fog deaths. During the autumn of 1909, in Glasgow, Scotland, 1,063 deaths were attributed to smoke and fog. In 1930, in Meuse valley of Belgium, a large number of people fell ill and many were killed by smog. The 'black fog' that settled on London in December 1952 caused 3,500 to 4,000 deaths. It was accompanied by thermal inversion over most of the British Isles. In Pennsylvania smog killed 20 people and about 40 per cent population became ill. The residents of the Kanto Plains area of Yokohama, Japan, were afflicted with an accumulation of smog during winters of 1945 and 1946. Similar type of attack was reported in New Orleans, Louisiana, in 1959. In recent years, with the increase in atmospheric pollution, the instances of smog resulting in serious injury or deaths have also become common.

The most irritating and injurious components of smog are the products of reactions in the atmosphere between oxygen, ozone and emission pollutants. The mixture of these products is called photochemical smog. These are reactions in which oxygen, ozone, nitrogen oxides and hydrocarbons produce those compounds which are toxic and irritating.

The constituents of smog are quite toxic and are responsible for respiratory and cardiac difficulties. Eye irritation is the most common symptom of smog injury. Similarly, bronchial asthma, chronic bronchitis and pulmonary emphysema are associated with smog. It is also injurious to animals and plants and one of the main causes of environmental degradation.

Pollution by Vehicular Exhaust

The automobile is man's greatest achievement in minimising distances. The number of automobiles is increasing day by day and has become a cause of air pollution and degradation of the environment. The automobile, with its internal combustion engine, emits poisonous gases, which are harmful to human health and is the most serious pollution problem of this technological age.

Exhaust emissions from diesel engines include carbon monoxide, hydrocarbons, sulfur oxides, organic acids, etc. The two primary pollutants are carbon monoxide and nitrogen dioxide, both of which are extremely poisonous gases and are present in the atmosphere of urban areas, specially in metropolitan areas. A third form of pollutant is constituted by the unburned hydrocarbons of gasoline. Lead is also a very toxic compound and its main source in the environment is believed to be from leaded gasoline used as fuel for internal combustion engines. The lead usually added to gasoline is the organic compound tetraethyl lead (TEL), which is extremely poisonous. The presence of lead in the atmosphere is a threat to the environment as well as for all living organisms.

It becomes clear from the above analysis that the problem of air pollution is increasing with the growth and expansion of industries and automobiles. It is high time that all of us should know the harmful effects of air pollution and also evolve technology to control it.

Effects of Air Pollution

Air pollution has now become a worldwide phenomenon and every individual in one way or the other is facing problems caused by it. Its impact can be seen locally, at regional level as well as at global level. At local and regional levels its manifestations are in the form of alterations in (i) visibility, (ii) intensity of sunshine, (iii) precipitation amount, and (iv) acid rain. Its global effects are: (i) change in natural climate by rise of temperature, melting of snow, (ii) increase in carbon dioxide, (iii) increase in particulates, (iv) holes in ozone layer, etc. Several aspects of air pollution, such as effects on the ozone layer, greenhouse effect, smog and acid rain have already been discussed. The effects of air pollution can be grouped under the following heads:

(i) effects on human health,
(ii) effects on animals and plants,
(iii) effects on atmosphere, and
(iv) other effects.

Effects on Human Health

Some environmental poisons can cause acute illness and even death. Others may be harmful, but the disease may take years or even decades to appear. Air pollution mainly affects the respiratory system. Bronchitis, emphysema, asthma and lung cancer are some of the chronic diseases caused due to exposure to polluted air. It is feared that lung cancer is caused mainly due to polluted air because carcinogens are found in it. Its mortality rate is higher in urban areas. Figure 7.1 shows the various effects of air pollution on the human body. Sulfur dioxide is the most serious and widespread air pollutant. Its lower concentration is a cause of spasms in the smooth muscle of bronchioles and its higher concentration induces increased mucus production. Sulfur dioxide is also considered to cause cough, shortness of breath, spasm of the larynx and acute irritation to the membranes of the eyes. It also acts as an allergenic agent. When it reacts with some compounds, sulfuric acid is formed which may damage lungs.

Figure 7.1

Effect of Air Pullution on the Human Body

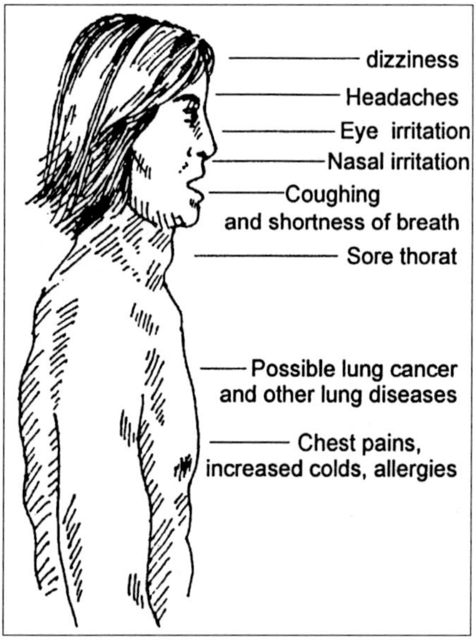

Carbon monoxide often affects the oxygen carrying capacity of blood. Nitric oxide is reported to be a pulmonary irritant and its excess concentration may cause pulmonary haemorrhage. Hydrogen sulfide is also toxic. Lead emitted from automobile exhausts is a cumulative poison and is dangerous particularly to children and may cause brain damage.

The particulate pollutants such as asbestos, silica, carbon, beryllium, lead, etc., are capable of exerting a noxious (fibrotic) local action in the interstitial areas of the lungs. Radioactive elements are also harmful to man and other living organisms. As described earlier, smog has a killer effect, which is also the result of air pollution. The death toll by smog varies from few persons to thousands. In December 1952, about 4,000 persons died in London due to smog. Similar cases have been reported from London itself in 1956, 1957 and 1962 in which the death toll was between 700 and 1,000 persons. In other countries also smog deaths have been reported. In fact, the growing air pollution has now become a health hazard for man. The main pollutants of air and their impact on health are shown below in Table 7.4:

Table 7.4

Main Pollutants of Air and their Impact on Health

Pollutant	*Sources*	*Health Hazards*
Carbon monoxide	Automobiles-due to incomplete combustion/ burning of carbon in the fuel of internal combustion engines.	Displaces oxygen in blood, reduces amount of oxygen to blood tissues. Dulls mental performance, causes accident proneness, extra burden on anaemic, heart and lung patients.
Particulates (soot, smoke, fly ash, etc.)	Stationary fuel combustion in industrial processes.	Occurrence of photochemical smog which can cause widespread deaths in a few days.
Oxides of sulfur	Mainly industrial processes involving burning of coal and oil as fuel.	Temporary or permanent injury to the respiratory system, irritate the upper respiratory tract and lung tissue.
Hydrocarbons	Automobiles, fuel combustion in electric power plants, industrial processes, combustion in solid waste disposal and agricultural burning.	Gaseous hydrocarbons in atmosphere are not toxic but help in forming photochemical smog.
Oxides of nitrogen	Burning of fuel at very high temperatures as in transport vehicles, power plants and industrial boilers.	Nitrogen oxides combine with hydrocarbons in the presence of ultraviolet rays of the sun to form secondary pollutants called photochemical oxidants.
Others	Photochemical oxidants.	Photochemical oxidants feroxy-acetyl nitrogen, aldehydes and acrolein) cause eye irritation and flu, when they combine with particulates to form smog.

Effects on Animals and Plants

The impact of air pollution on animals is more or less similar to that on man. Chronic poisoning results from the ingestion of forage contaminated with atmospheric pollutants. Among the metallic contaminants, arsenic, lead and molybdenum are important. Fluoride is another pollutant, which causes fluorosis among animals. A number of livestock have been poisoned by fluorides and arsenic in North America. Bone lesions in animals due to excessive fluorides have also been reported.

Air pollution has caused widespread damage to trees, fruits, vegetables, flowers and in general, vegetation as a whole. The total annual cost of plant damage caused by air pollution in USA alone has been estimated to be in the range of 1 to 2 billion dollars. The most dramatic early instances of plant damage were seen in the total destruction of vegetation by sulfur dioxide in the areas surrounding smelters. When the absorption of sulfur dioxide exceeds a particular level, the cells become inactive and are killed, resulting in tissue collapse and drying of leaves. Cotton, wheat, barley and apple are more sensitive to this pollutant.

Fluorides are responsible for various types of injuries to plants. The leaves of apple, apricot, fig, peach and prune are more susceptible to air borne fluorides. Fluorides seem to interfere with the photosynthesis and respiration of plants. Smog also causes injury to plants. Similar impact of ozone can be seen in the lesions to plants. Chlorine, ammonia, hydrogen sulfide, etc., are also harmful to vegetation.

Effects on Atmosphere

Some of the effects of air pollution on atmospheric conditions, such as effect on ozone layer, greenhouse effect, etc., have already been discussed. There is an increase in the carbon dioxide concentration in the air due to increased combustion of fossil fuels. Carbon dioxide absorbs heat strongly and the radiative cooling effect of the earth is thus decreased. The rising of temperatures and ozone holes are some of the problems which have attracted the attention of the scientists all over the world. These problems are not related to any region or a country but are the global problems and their impact on world climate may be hazardous to the whole world. The local weather conditions are highly susceptible to air pollution. Its impact on temperature, humidity, rainfall and clouds is apparent. The 'smog dom' on large urban centres is the result of air pollution. Due to air pollution, visibility also reduces.

Other Effects

Air pollution can also cause damage to property and materials. The smoke, grit, dust and oxides of sulphur have harmful effects on structures. In 1972, when an oil refinery at Mathura was opened, its impact on Taj Mahal became a major issue. Sulfur dioxide is the most damaging of gaseous pollutants. Aluminum alloys, copper and copper alloys, iron and steel are corroded when exposed to contaminated air. Hydrogen sulfide reacts with lead paints to form lead sulfide thereby producing a brown to black discolouration. The damage caused by air pollution to structures is not serious but from an aesthetic point of view, it is not desirable.

Management Measures

Worldwide efforts are going on to control air pollution. Scientists and technologists have developed certain measures but still progress in this field is not satisfactory. For keeping air pollution under control, there is a need to deal this problem from two angles: (i) to check the present prevalent air pollution in such a way that its harmful effects can be minimised, and (ii) to develop such methods and technology so that air pollution can be controlled in future. Some of the measures to contain air pollution are as follows:

1. The forest cover should be protected by restricting deforestation and through adoption of afforestation programmes. Trees are the best controllers of air pollution. It is an accepted fact that alt least 33 per cent of the land area should remain under forest/tree cover. It will help in containing air pollution and also help in maintaining the ecological balance.
2. There must be a 'green belt' around every township and village. Similarly, industrial areas should be surrounded by green belts. Table 7.5 indicates the width of green belt for various industries.

Table 7.5
Width of Green Belt for Various Industries

Industry	*Examples*	*Area of Works (ha)*	*Width of Green Belt*
Heavy industry	Oil refineries, chemical works, metallurgical and seaport industries, nuclear reactors	>500	<2 km
Heavy industry	Machine manufacture, ship-building, big harbour industries, power stations	200–500	> 1 km
Medium-heavy industry with much air pollution	Manufacture of straw boards, artificial fibres, and ceramic products, cement works	100–200	500 m or more
Medium-heavy industry with little air pollution	Manufacture of cars, foods and textiles		200 m or more
Light industry with some air pollution	Tannerie, textiles and food industries	50–100	50–100 m
Light industry with some air pollution	Manufacture of electronic apparatus and domestic machines		
Service industry	Printing works, bakeries, laboratories	10–50	>100 m
Workshops, handicrafts, etc.	Fashion studios, photoprinting shops, potteries	1–10	>50m

Adapted from *Manual on Urban Air Quality*, WHO Regional Publications, European Series No.1, Copenhagen, 1976.

3. The main source of air pollution is the automobiles, therefore, their engines should be redesigned in such a way that their emissions cause minimum pollution. Several steps have been taken and some technology has also been developed, but still it is in a very preliminary stage. Apart from technological changes, some steps on individual level should be taken by vehicle owners, such as:

(i) regular service of the vehicle,
(ii) change or replacement of old engines,
(iii) use of filters and after burner,
(iv) engine should be well tuned, and
(v) mixing of anti-tibs in diesel and petrol, etc.

4. The use of steam engine by the railways should be stopped or minimised. As far as possible, electrically operated rail engines should be used, which are pollution free.
5. In countries like India, traditional use of fuelwood, coal, etc., should be checked and newly devised smoke-free furnaces be used.
6. The industrial areas should be located at a certain distance from residential areas. The selection of the site should be done with consideration of air direction.
7. Steps should be taken for immediate check on forest fires.
8. In industries, arrangements for pollution control should be done. Only after full arrangement of effluent emission, permission for production should be accorded.
9. With the help of chemical reactions harmful effects of pollution can be controlled.
10. Some methods of controlling air pollution are filtering, settling, dissolving, absorption, etc. For these methods cheap devices should be developed.

Actual prevention of air pollution is not so simple as it looks. However, it can be minimised or controlled without undue cost by careful planning. For this, the following steps will be helpful:

Control of Gaseous Pollutants

The equipments which can be used to control gaseous pollutants are classified as combustion, absorption and adsorption equipments. *Combustion* is applicable to pollutant gases which are oxidizable. In petrochemical, fertiliser, paint and varnish industries combustion control equipments are useful. *Absorption* is a diffusional process in which the transfer of gas molecules into a liquid phase takes place. Such equipments include spray chambers, packed towers and sieve plate contractors. Absorption is the means of controlling air pollution which occurs due to some gases and vapours and inflammable compounds which cannot be treated by other means.

Control of Emission from Motor Vehicles

The emission control techniques include tune-ups, catalytic reactors and engine modifications. A high air-fuel ratio will reduce the concentrations of both carbon monoxide and hydrocarbons. The modified engines have an efficient system of burning of the fuel. At present, there is no system of reduction of sulphur dioxide, although researches in this direction are in progress throughout the world.

Control of Aerosol Emissions

This can be controlled by arresters and scrubbers. Arresters include inertial separators, filters and precipitators. The electrostatic precipitator is considered to be the most effective device for preventing the emission of dust from fuel gases and is the standard equipment for large power stations.

Control by Fuel Selection and Utilisation

Coal and oil are the primary fuels in which smoke, grit and sulfur dioxide are the major pollutants. Coal pollutes more because it releases more smoke and carbon in the air. Instead of coal, oil can be used. But, in oil, the amount of sulphur dioxide emission is higher. Therefore, fuel selection should be done properly and the harmful impact be restricted by inducing other chemicals. In order to control smoke, the coal is pulverised before being used. When oil is used for fuel, it is essential that the ratio of air should be maintained constant in order to prevent smoke emissions.

Control of Air Pollution by Site Selection and Zoning

Selection of an industrial site is the most important factor through which impact of air pollution can be minimised. The industrial site should be selected considering: (i) residential areas, (ii) nature of industries, and (iii) climatic conditions, specially the direction of winds. The site of each and every industry should have an approval from the department of environment. It should be approved in taking into consideration the long term perspective rather than immediate effects. Within the city the location of industrial zone is a critical aspect and town planners should have a check over its location as well as unorganised growth.

Apart from the above, legal controls are also there. The UNO and WHO have issued guidelines to its member-nations for the formulation of laws regarding prevention of air pollution and accordingly all the countries of the world have enacted certain laws. In India, the Prevention of Air and Water Pollution Act, 1974, 1981, the Air Prevention and Control of Pollution Act, 1981, and the Environmental Protection Act, 1986 have been enacted, but they are not followed in a strict manner. Actually, the most important aspect is public awareness about air pollution because the government alone cannot prevent it. It can be checked only through the combined efforts of the government, NGOs and the public.

References

Khemani, L.T. et al., spread of Acid Rain over India, *Atmospheric Environment*, 1989, Vol. 23, pp. 757–62.

Maxwell, K.E., *Environment of Life*, Dickenson Pub. Com., California, 1973, p. 128.

Strahler, A.N. and Strahler, A.H., *Geography of Man's Environment*, John Wiley, 1977, p. 54.

8

Water Pollution and Its Management

Water is essential for life on earth. We drink it, bathe in it, relax in it, fish in it, keep cool with it, irrigate the plants, produce energy with it and also use it for transportation and recreation. Many ancient civilisations have developed in river valleys like the Nile, Indus, Tigris and Euphrates, etc. Even today water provides a base for social, cultural and economic development. It seems that water is abundant, but usable water is very limited and creates a serious conservation problem in many places where it is needed. Apart from scarcity of water, we are also facing a problem of water pollution not only of rivers and lakes but also of underground water. The intolerable burden of chemical and human waste products has become a threat to aquatic as well as terrestrial life. Therefore, much attention has been given to study the various facets of water pollution, not only to understand the nature and effects of water pollution but also for its control.

Sources of Water and Its Quality

The planet earth has 71 per cent of its area as oceans and the oceans contain 97 per cent of all water in the form not suitable for human consumption. Much of the remainder is frozen in the ice caps and glaciers. The water in rivers and lakes is very small. Less than 1 per cent is in the form of ice-free fresh water, yet this negligible portion of the planet's water is crucially important to all forms of terrestrial and aquatic life. Apart from surface water available in rivers and lakes, underground water resources are also important. The percentage of land water distribution over earth's surface (other than oceans) is as follows:

Rivers and streams	00.014%
Fresh water lakes	01.4%
Salt lakes and inland seas	01.2%
Soil moisture and seepage	00.7%
Groundwater (one km deep)	48.3%
Groundwater (below one km deep)	48.3%

Fresh water is a scarce commodity with greatest amount locked in glaciers and ice caps. But it is constantly replenished in a cyclic way known as hydrological cycle. The hydrological cycle is a simplified description of water movement from place to place. This water cycle pattern has been depicted in Figure 8.1. As shown, the cycle begins with evaporation by solar energy. The formation of clouds (condensation) leads to subsequent precipitation which results in the conversion of salt water to fresh water. The precipitation on the surface may run off as rivers or infiltrate into the soil and rock and form groundwater. Surface and sub-surface water returns to the ocean unless evaporation interrupts the cycle.

The quality of water is the most important factor. From chemical point of view, water is the monoxide of hydrogen (H_2O) in which two parts of hydrogen and one part oxygen are present. But, this form of pure water is not available in nature. As soon as water vapour enters into the atmosphere after condensation, impurities like dust particles, gases, etc., are mixed in it and when it reaches the surface, several minerals and chemicals are also mixed in it. The quality of drinking water depends on the quantity of harmful elements present in it. Drinking water should be clear, odourless and tasteless and its pH value should be between 7 and 8.5. In 1971, WHO has given the following limits of the various impurities in drinking water:

Figure 8.1
The Hydrological Cycle

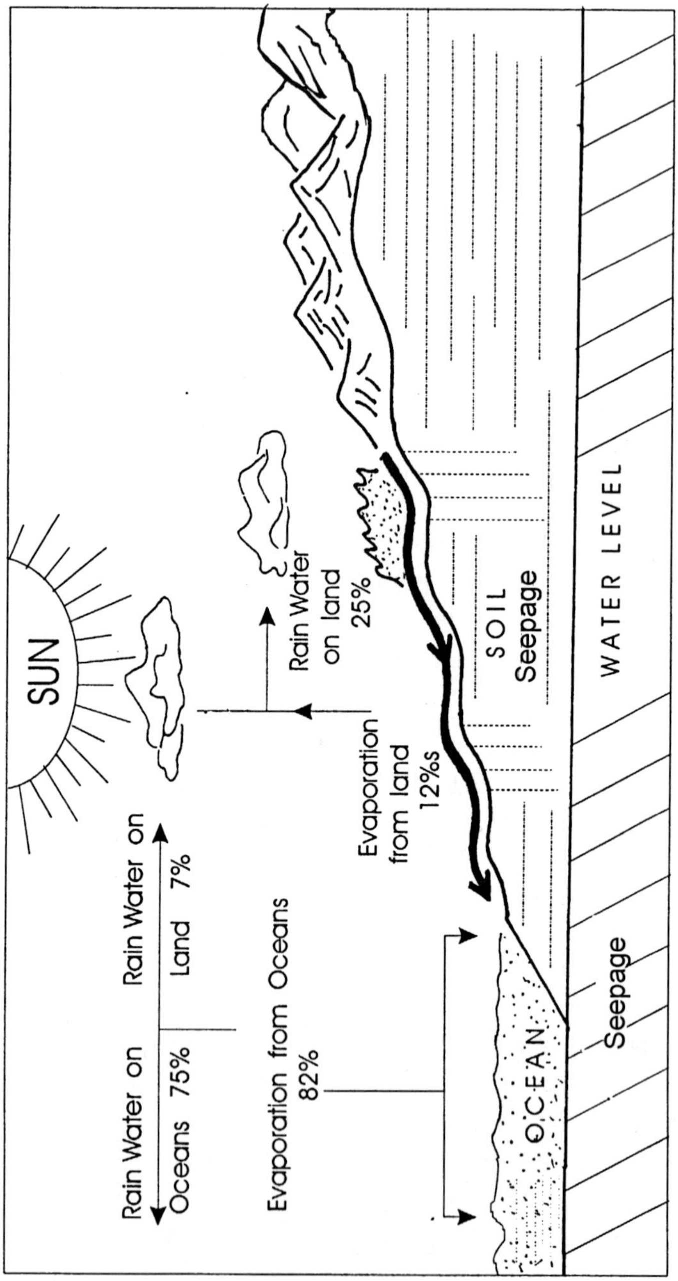

Table 8.1
Permissible Limits of Impurities in Water

Impurities	*Max. Permissible Limit (mg/litre)*
Total solid matters	500
Hardness	2 mEg/Litre
Calcium (Ca)	75.00
Magnesium (Mg)	30.00
Sulfate (SO_4)	200.00
Chlorides	200.00
Iron (Fe)	0.10
Manganese (Mn)	0.05
Copper (Cu)	0.05
Zinc (Zn)	5.00
Arsenic (As)	0.05
Cadmium (Cd)	0.005
Cyanide (Cn)	0.05
Lead (Pb)	0.05
Mercury (Hg)	0.001
Phenolic elements	0.001

The permissible organic impurities include Bacillus coli not more than one in 100 m/litre and other Coli bacteria not more than 10. In natural water impurities are found in three forms:

(i) suspended impurities,
(ii) colloidal impurities, and
(iii) dissolved impurities.

Suspended impurities float on the water surface as dust particles, particles of minerals, paper, cloth, foam, leaves and floating vegetation as well as other organic and inorganic materials. Colloidal impurities are those which are mixed with water and difficult to separate, such as silica glass, oxides of various minerals, micro bacteria, etc. On the other hand, dissolved impurities become part of the water during its flow. These include carbon dioxide (CO_2), sulfur dioxide (SO_2), methane (CH_4), sodium (Na^+), potassium (K^+), calcium (Ca^{2+}), magnesium (Mg^{2+}), iron (Fe^{2+}), ammonium (NH_4^+), etc. Apart from these, man has induced several impurities in the water, which are responsible for water pollution.

Water Pollution

Water pollution simply means contamination of water due to any external material, or, in other words, introduction of something to natural water which makes it unsuitable for human consumption. WHO has defined water pollution as "any foreign material either from natural or other sources that may contaminate the water supply and makes it harmful to life, cause of their toxicity, leads to reduction of normal oxygen level of water, causes aesthetically unpalatable effects and spread of epidemic diseases." Owen (1971) has defined it as "any unreasonable contamination of water which lessens its value to man." In general, "water pollution may be defined as the adverse change in composition or condition of the water such that it becomes less suitable for the purposes for which it would be suitable in its natural state. The changes include physical changes, chemical changes and biological changes" (Manivasakam, 1992). According to Gilpil (1978), "the deterioration in chemical, physical and biological properties of water brought about mainly by human activities" is the water pollution. Whatever additions are done to the water, either by natural processes or by human activities invariably, they change the natural qualities. The problem of water pollution was first recognised by Hippocrates (450 BC), who suggested filtration and boiling as remedial measures. With the fast increase in our industrial civilisation, the demand for water is also increasing day by day. At the same time, population explosion, urbanisation, improper sewage disposal, unsafe industrial wastes, radioactive wastes, oil leakages, etc., have polluted our water resources so much so that about 70 per cent rivers and streams not only of India but of all the countries contain polluted water.

Sources of Water Pollution

Water pollution is caused by several sources, which are not independent in nature but interact with one another. Generally, one or two factors become prominent and may be considered the primary source of water pollution. The various sources of water pollution can be classified into natural and human as shown below:

Figure 8.2
Sources of Water Pollution

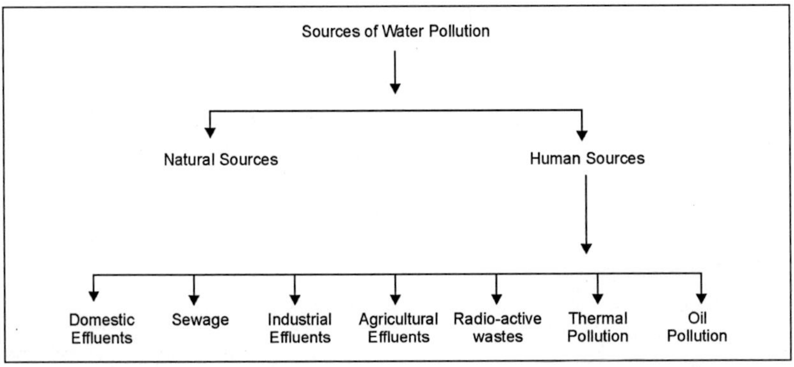

Natural Sources

Some natural elements, like gases, soil, minerals, humus material, waste created by animals and other living organisms present in the water, creat water pollution. During rains, one can see the direct impact of soil on the water of rivers, tanks and other water bodies. The brown and dirty water is the result of mud mixed in the water, which becomes clear after some time. Similarly, several types of natural suspended and colloidal impurities are also there in the water. These elements may be organic or inorganic and may even be harmful. The minerals like sodium, potassium, calcium, magnesium, iron, etc., are also present in the water, and if their quantity is more than the permissible limit, these can be harmful. Some poisonous minerals, such as nickel, beryllium, cobalt, lead, mercury, cadmium, etc., are very harmful and are responsible for water pollution. Sometimes water-borne vegetation also becomes a cause of water pollution.

Human Sources

There is no doubt that water pollution is mainly the result of human activities. The rapid growth of population, urbanisation, industrialisation and increasing use of chemicals have all resulted in water pollution and this problem is increasing day by day in spite of several measures taken in this direction. This is not only a problem of developed countries but also developing countries. Both urban as well as rural areas are affected by it. A misconcept developed right from the dawn of human civilisation is that water bodies can be used as dumping grounds for all types of wastes and this is responsible for water pollution throughout the world. This is not only hazardous to human health and other organisms but also disturbs the delicate ecosystem. Table 8.2 indicates sources or types of contaminates responsible for water pollution and their impact.

For detailed analysis, the human sources of water pollution can be classified into the following categories:

(i) domestic effluents and sewage,
(ii) industrial effluents,
(iii) agricultural effluents,
(iv) radioactive wastes,
(v) thermal pollution, and
(vi) oil pollution.

(i) Domestic Effluents and Sewage

Man, for his various domestic purposes such as drinking, cooking, bathing, cleaning, cooling, etc., uses on an average 135 litres of water per day. About 70 to 80 per cent of this is discharged and drained out, which through municipal drains poured into, in many cases, a river, tank or lake. This water is known as domestic waste water and, when other waste material such as paper, plastic, detergents, cloth, etc., is mixed in it, it becomes municipal waste or sewage. The domestic waste water and sewage is the main source of the water pollution. This is the inevitable and unfortunate fallout of urbanisation. This organic waste depletes the oxygen from water and upsets the natural balance of the aquatic ecosystem.

Municipal sewage is considered to be the main pollutant of water. Most of the sewage receives no treatment before discharge, specially in developing countries like India. In Delhi alone, 120 crore litres of water is consumed per day, out of which 96 crore litres is drained into the Yamuna river through 17 big drains. In the same manner, all the 47 towns located on the bank of river Ganga

Table 8.2
Water Pollution Contaminants and their Impact

Source or Type of Contaminant	*Impact*
1. Degradable Wastes	
Domestic and municipal sewage and other oxygen demanding industrial wastes—they are acted upon by the bacteria and oxygen dissolved in water and reduced to inorganic form, as also in quantity.	Depletion of dissolved oxygen in water—harmful for fish and other aquatic life.
2. Plant Nutrients	
Phosphatic and nitrogenous compounds dissolved in water from industrial wastes, or due to washing away of fertilisers due to heavy rains from fields.	Stimulate growth of algae/plankton and other plants resulting in clogging of waterways and rendering the water unfit for human use.
3. Infectious Agents	
Polluted, stagnant water breeding grounds for parasites, bacteria, viruses of all kinds.	Cause water-borne diseases and outbreak of epidemics such as ameobiosis, dysentery, cholera, typhoid, etc.
4. Synthetic Organic Compounds	
Such as pesticides, agricultural chemicals, detergents, industrial wastes, DDT, etc.	Instance of cadmium, lead and mercury poisoning in human beings. Cause neurological impairment and even death.
5. Inorganic Minerals and Other Chemicals	
Asbestos and acids, etc.	Asbestos produces lung cancer. Acids cause allergies, ulcers, skin diseases. Also have catastrophic impact on fish and aquatic life.
6. Radioactive Elements	
Wastes from nuclear power plants and nuclear fuel reprocessing plants are packaged and buried under water/ oceans.	Potentially hazardous in the event of leakages and induce radiation-related illness.
7. Sediments	
Soil and mineral particles washed into streams or water.	Reduce amount of sun-light available for marine plants, cause clogging of filter plants.
8. Thermal Pollution	
Water used for cooling in electric, thermal, nuclear power plants, when recirculated in water bodies.	Can raise lake water temperature by 70-100°C aggravate diminution of dissolved oxygen, kills marine fish and plant life.

drain their sewage into it. With the growth of population, the quantity of waste water is also increasing in addition to the production of large quantities of sewage. Sewage contains decomposable organic matter and exert an oxygen demand on the receiving waters. The common organic materials found in a sewage are soaps, synthetic detergents, fatty acids, and proteinaceous matters such as amines, amino acids, amides and amino sugars. Besides, it also contains numerous micro-organisms in the form of pathogenic bacteria and viruses derived from human faces. Untreated waste water is often the carrier of viruses and bacteria and, with poor household sanitation practices, has been linked to high infant mortality rates in developing countries. Even where most sewage is treated, as in the developed world, recent measurements of fecal coliforms in some countries indicate increasing pollution. Sewage supports the growth of other forms of life that consume oxygen, it is measured in terms of Biochemical Oxygen Demand (BOD). It is the lack of oxygen that kills fish and other aquatic life.

In recent years, there has been considerable growth in the use of detergents, which causes severe water pollution. Many modern detergents contain phosphates, which are an essential component of agricultural fertilisers. When phosphate detergents are discharged into waterways, they supply a needed nutrient and promote rapid growth of algae. This enrichment process is known as 'eutrophication'. In many areas of the world, aquatic weeds have multiplied explosively. They have interfered with fishing, navigation, irrigation and even production of hydroelectricity. In developing countries, human population and settlements are growing fast, often faster than waste water treatment facilities can be provided. Thus, much of the untreated waste water and sewage is discharged into rivers and other water bodies, making the water unsuitable for drinking.

(ii) Industrial Effluents

Industrial activities generate a wide variety of waste products, which are normally discharged into water courses. Major contributors are the pulp and paper, chemicals, petrochemicals and refining, metal working, food processing, textile, distillery, etc. The wastes, broadly categorised as heavy metals or synthetic organic compounds, reach bodies of water either through direct discharge or by leaching from waste dumps. In developed countries of the world, many industrial discharges are strictly controlled. Yet, water pollution continues from accumulations of wastes discharged over the past 100 years. But, in developing countries, industrial discharges are largely uncontrolled and thus, a major cause of water pollution.

All the Indian rivers have been polluted by industrial effluents. The 'holy' river Ganga has become a highly polluted river today due to various types of industrial discharges. Along the Ganga, several chemical, textile, tanning, pulp and paper, petrochemical, rubber, fertiliser and other industries are located and all of them discharge their waste water and other effluents, directly or indirectly, into the river, resulting in the pollution to such an extent that even the Ganga Action Plan, to control water pollution, has failed miserably. From Delhi industrial area alone, more than 8 lakh tonnes of industrial waste is discharged into the river Yamuna.

Damodar river of Bihar is a highly polluted river due to industrial wastes discharged from Bokaro, Rourkela, Indian Iron and Steel Company (IISCO), Bengal Paper Mills, Sindhri Fertiliser Factory, etc. A study reveals that from Durgapur Steel Plant 1,800 cu/m washed coal has been discharged into the river. Similarly, from IISCO 15,000 cu/m and from Bengal Paper Mills 12,000 cu/m industrial waste is discharged per day into river water. Likewise, about 10 to 15 tonnes of sulfuric acid from Sindri and 5 to 10 tonnes toxic chemicals are discharged into the river water. The story of the Hooghly river in West Bengal is more or less same. Its water has been polluted to such an extent that even fish fertilisation has become difficult. The paper mills located along the river discharge about 11.4 tonnes liquid wastes into the river. Almost all other rivers of India have the same fate. Chambal, Narmada, Kaveri, Godavari, Mahanadi and all other small rivers have been polluted and if this is not stopped, it will result in greater water pollution.

The nature and effects of the pollutants from the effluents of paper and pulp, textile, food processing, chemicals, metal and petroleum industries are as follows:

- Effluents from *paper and pulp industries* include wood chips, bits of bark, cellulose fibres and dissolved lignin, in addition to a mixture of chemicals. All these produce a sludge which blankets fish spawning grounds and destroys certain types of aquatic life. The effluents contain chlorine, sulfur dioxide, methyl mercaptan, etc., which are considered to be highly poisonous to fish.
- Effluents from *textile industries* are alkaline in nature and have a higher demand for oxygen.
- *Food processing industries* include dairies, breweries, distilleries, meat-packing, etc., where the waste products include fats, proteins and organic wastes. These industries discharge wastes containing nitrogen, sugar, proteins, etc. This waste contains a higher BOD and is therefore responsible for water pollution.
- *Chemical industries* include acid manufacturing, alkali manufacturing, fertiliser, pesticides and several other industries. The effluents from these industries contain acids which have corrosive effects. The effluents from fertiliser industries contain phosphorus, fluorine, silica, and a large amount of suspended solids.
- *Metal industries* usually discharge effluents containing copper, lead, chromium, cadmium, zinc, etc., which are toxic to man as well to aquatic life. These wastes also contain acids, oils, greases and cleansing agents.
- *Petroleum industries* include oil refineries and petrochemical plants. The effluents include hydrocarbons, phenolic compounds and other organic and inorganic sulphur compounds.
- *Other industries*, which pollute water are tanneries, soaps and detergent industries, glass, electroplating, bleaching, atomic plants, explosive factories, etc.
- *Mining operations* can result in metals leaching into the acidic effluents, thus adding to the metal load in rivers, lakes and groundwater. Discharge of mercury from gold mining activities has polluted some streams in Brazil and Ecuador and created serious health problems. With reference to water pollution through mercury, mention of Minimata Gulf

incident must be made. In 1950, near the Japanese coast, in Minimata Gulf, fishermen suffered from blindness, weakness, mental illness, paralysis, etc. It was found that effluents discharged from a plastic factory contained mercury, which entered the fish and by eating those fish, all the fishermen suffered from effects of mercury poisoning. Thus water pollution through industrial effluents has become a major environmental problem today and needs measures to control it.

(iii) Agricultural Effluents

Agricultural water pollution is caused by fertilisers, insecticides and pesticides, farm animal wastes and sediments. In recent years, use of chemical fertilisers has increased manifold. The green revolution in India is a reflection of the increased use of fertilisers. The chemicals used in fertilisers enter the groundwater by leaching and the surface waters by run-off. The nitrates, when mixed with water, may cause methemoglobinemia in infants. Incidences of nitrate poisoning are also seen in livestock. The plant nutrients, nitrogen and phosphorus are reported to stimulate the growth of algae and other aquatic plants.

The use of various types of pesticides and insecticides in agriculture is also one of the causes of water pollution. Their presence in water is highly toxic to both man and animals, because all these have a high persistence capacity, i.e., their residues remain for long periods.

The farm animal wastes often pose serious problems of odour and water pollution. These wastes also contain pathogenic organisms which get transmitted to humans. Sediments of soil and mineral particles washed out from fields also cause water pollution. They fill stream channels and reservoirs and reduce the sunlight available to aquatic plants.

(iv) Radioactive Wastes

Radioactive elements, such as uranium and radium, possess highly unstable atomic nuclei. This disintegration results in radiation emission which may be highly injurious. During nuclear tests, radioactive dust may encircle the globe at altitudes of 3,000 metres or more, which often comes down to the earth as rain. Eventually, some of the radioactive material, such as Strontium 90 (which can cause bone cancer), percolates down through the soil into groundwater reservoirs or is carried out into streams and rivers. In both cases, public water supplies may be contaminated. The construction of more nuclear reactors and the increasing use of radioactive materials in medical research represent other potential contamination sources.

(v) Thermal Pollution

Most of the thermal and electric power plants also discharge considerable quantities (about 66%) of hot effluent/water into nearby streams or rivers. This has resulted in thermal pollution of our water courses. Thermal pollution is undesirable for several reasons. Warm water does not have the same oxygen holding capacity as cold water. Therefore, fishes like black bass, trout and walleyes, etc., which require a minimal oxygen concentration of about 4 ppm, would either have to emigrate from the polluted area or die in large numbers. When the temperature of the receiving water is raised, the dissolved oxygen level decreases and the demand for oxygen increases, hence anaerobic conditions will set in resulting in the release of foul gases. Thermal pollution is considered hazardous for the whole aquatic ecosystem. Several industries have installed cooling towers, where the heated water is cooled. But even so, thermal pollution has become a serious problem for water bodies located near thermal plants.

(vi) Oil Pollution

The spread of oil in the sea has become a common feature nowadays. Oil is transported across oceans through tankers and either due to some accident or leakage oil spills onto the water and causes the degradation of aquatic and marine environment. Between 1968 and 1983, there were more than 500 tanker accidents that involved oil spills. Altogether, more than one million tonnes of oil was released. A dramatic incident was that of the tanker *Torrey Canyon,* when it struck off the southern tip of the British

Isles in March 1967. The *Torry Canyon* was the largest oil spill up to that time. The pollution caused widespread destruction of many forms of marine life despite strenuous efforts to clean up the spill. Similarly, on March 16, 1978, the oil tanker *Amoco Cadiz,* lost its steering off the coast of Brittanny in France and the total spilling of oil was 1.6 million barrels. Such accidents have become very common due to technical problems or heavy marine traffic. During the 1991 Gulf War, there was heavy bombing on oil tanks, which resulted in the spilling of oil. The impact of this oil spill on the marine ecosystem in this area has not yet been remedied. Offshore drilling operations also contribute their share of oil to the sea. The total quantity of oil that finds its way into sea each year is very large. It has been estimated that about one million tonnes of oil spills into the ocean each year from tankers and oil drilling operations.

Harmful Effects of Water Pollution

Since use of water is universal to all living organisms, the impact of its pollution is also widespread. The effect of water pollution has become severe because water is the controlling factor in the 'food chain', therefore, its impact is not only widespread but intensive also. The harmful effects of water pollution can be divided into three groups, viz., (i) effect on man, (ii) effect on aquatic life, and (iii) other effects.

Effects on Man

On a worldwide scale, the pollution of water supplies is probably responsible for more human illness than any other environmental influence. The diseases so transmitted are chiefly caused by micro-organisms and parasites. Sewage and polluted water are responsible for several water-borne diseases. Some of the diseases caused by polluted water are cholera, typhoid, infantile diarrhoea, dysentery, infectious hepatitis, polio, giardiasis, amoebic dysentery, jaundice, zondic, etc. Cholerais, an illness caused by ingestion of the bacterium *Vibriocholeral,* results rapidly in massive fluid depletion and death in a very large percentage of patients. It is transmitted by drinking water contaminated with the faces of infected individuals. In developing countries, deaths by cholera are still common. Guineaworm is another disease caused by polluted drinking water. In India alone, about 18 lakh people suffer from this disease. Schistosomiasis is a group of diseases caused by infection through worms in water. There are certain chemicals present in polluted water which may cause diseases like nepurities, wrist drop, foot drop, etc. In spite of drastic measures taken for the supply of purified drinking water in USA, there are incidents of diseases caused by water contamination in one way or other. The developing countries of Africa, Asia and Latin America, where arrangements of water treatment are limited or negligible, a large section of the population is susceptible to water-borne diseases.

Effects on Aquatic Life

The effect of water pollution is more on aquatic life, because their existence depends on water and when there is any disturbance in their ecosystem, the impact is maximum on them. In polluted water, due to abundant growth of algae, the oxygen content becomes lesser, causing the death of fishes and other organisms. It is estimated that during the last twenty years, there is a decrease of about 40 per cent in aquatic life. Algae are responsible for imparting a peculiar taste and odour to many of the water supplies and also are a cause of gastroenteritis. A report by Wheler indicates that algal poisons usually act on the central nervous system and skin and is reported to be capable of producing cirrhosis of the liver.

There are many cases on record of the destruction of marine life by polluted water. Mass killing of fish was among the earliest and most dramatic result of indiscriminate pollution of water. Another type of fish killer is pollution from municipal sewage. Industrial, thermal, radioactive pollution, etc., are also threats to marine life. Because of water pollution large number of ducks also die every year. For example, 1,40,000 ducks died of botulism in California in 1970. Occasional deaths of animals have also been reported by drinking of polluted water. If water pollution is more severe, the process of photosynthesis is also obstructed which affects the growth of aquatic vegetation. All the chemicals that are drained into the water have harmful effects on every organism that survives there.

Other Effects

Due to water pollution, the physical and physiological nature of water has also changed. The colour of water changes due to pollutants, mainly due to organic dyes. The colour in itself is not harmful, however, aesthetic considerations make coloured water unsuitable for any purpose. The turbidity caused by the presence of colloidal or suspended matter is the characteristic trait of sewage and industrial effluents. Generally, the more the turbidity, the stronger is the sewage or effluent and the effects are worse. Foam is the product of water mixed with soaps and detergents. It consists of a suspension of air bubbles in water medium. Industrial effluents containing chemical substances, such as iron, chlorine, phenols, etc., affect the taste of the water. The decomposition of organic matter and algae, fungi and filamentous bacteria also imparts peculiar tastes. Similarly, the odour of water also changes due to the presence of pollutants. Even small amount of sulfides cause bad odour in water. The pollutants also reduce the transparency of water. Oil spills have their effect over evaporation. The industrial effluents, which contain several types of chemicals, are also responsible for the loss of soil fertility. In fact, the effects of water pollution are multidimensional in nature. Apart from above mentioned direct effects, there are several indirect effects also. Thus, water pollution has become a threat to the ecosystem and an important cause of environmental degradation.

Management of Water Pollution

As stated earlier, water pollution has become a worldwide problem today. Not only developing countries but developed countries are also facing this. However, developed countries have initiated several methods not only to control water pollution but also for the purification of polluted water. But, most of these techniques are expensive and even those countries are experiencing difficulties in containing water pollution. Apart from technological methods, there are other measures which can be useful in minimising water pollution. Some of them are as follows:

(i) All domestic and municipal effluents be drained in water bodies only after proper treatment.

(ii) As far as possible, domestic waste water can be used for irrigation as, it is very useful for horticulture.
(iii) Similarly, if sewage is mixed with soil, it will act as a fertiliser.
(iv) The drinking water sources and their nearby area should be kept clean and regular cleaning be done.
(v) There must be strict regulations for industrial effluents. In any case, it should not be discharged into water bodies. Every industry should develop its own effluent treatment plant.
(vi) Use of pesticides in agriculture should be minimised and only standard quality pesticides be used.
(vii) There is also a need to conserve several water purifying organisms.
(viii) In the Indian context, there should be a complete ban on the disposal of dead bodies in rivers. The bank of water bodies is often used as public latrines, this should be strictly checked.
(ix) Algae and other water-borne vegetation should be cleaned regularly.
(x) Chemicals, such as potassium permanganate, should be spread regularly in order to protect water from micro-organisms.
(xi) Under present practices, pure, clean, drinking water is used to flush toilets and irrigate lawns. There are two possibilities to check this loss—first, waste water from bath room and sinks be recycled and be used for irrigation and flushing of toilets and second, separate water supply system be provided for drinking water and water for other uses.
(xii) The most important aspect would be to educate people not to pollute water. This awareness will help in solving this problem to a great extent.

Treatment of Sewage

The treatment of sewage is one of the important measures, which aims in the removal of BOD, phosphorous, nitrogen, solids and bacteria. The composition of sewage is complex, and it differs depending upon the sources, the type of treatment or lack of it. The process of treating sewage is broadly classified as *primary, secondary* and *tertiary* (see Figures 8.3, 8.4 and 8.5).

Figure 8.3
Sewage Treatment Process

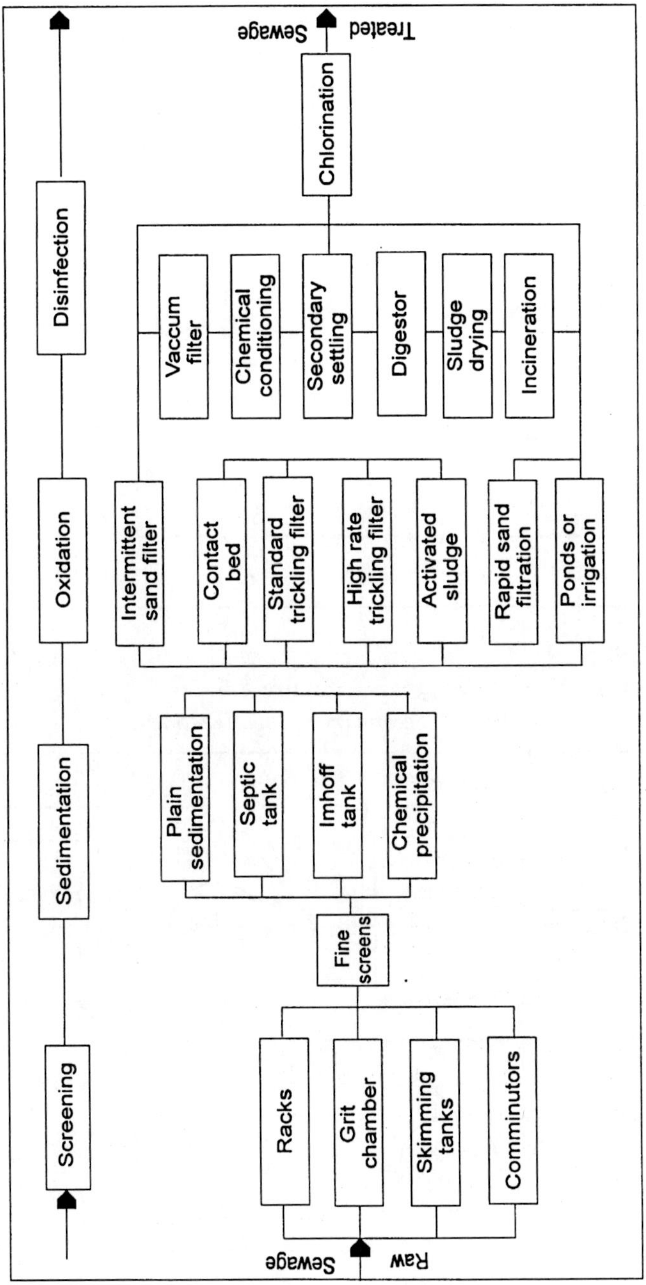

Figure 8.4
Primary Sewage Treatment

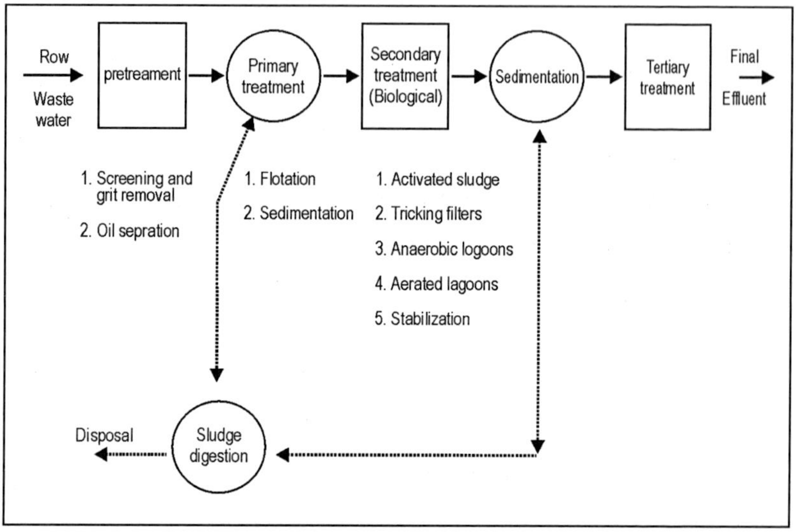

Figure 8.5
Secondary Sewage Treatment

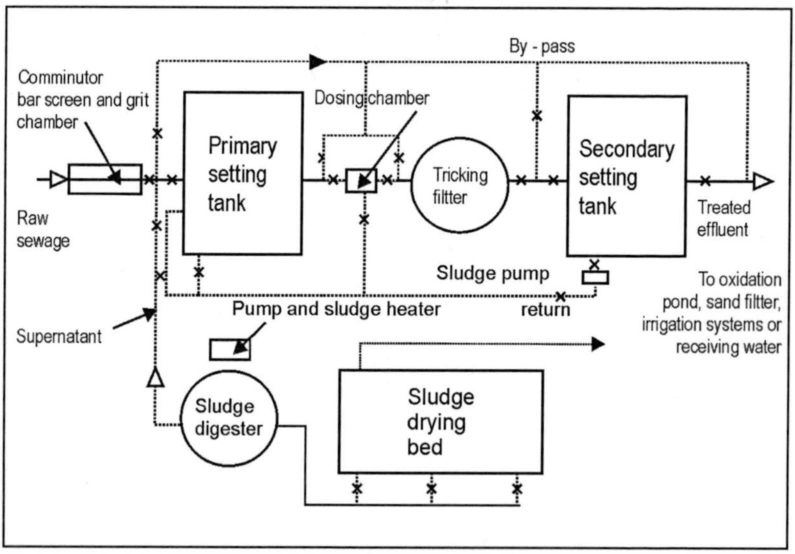

Primary Treatment

Primary treatment consists of removing floating and suspended solids by mechanical means. More than half of the suspended solids can be removed by primary treatment as shown in Figure 8.4.

In this process, first, the large solids are screened out and grease and scum are removed. Screening is done through metal bars first spaced 25 to 50 mm apart and may range down to 0.8 mm openings. Sand and other coarse material is removed by grit chambers. After screening and removal of grit, the waste water is run directly into settling or sedimentation tanks, the process by which the suspended solids are removed by gravitational setting. Primary sludge is a problem because it is bulky and must be removed. It also contains 94 to 99 per cent water. In some cases, sludge is dried in beds with some water removed by filtration. A report of the American Chemical Society indicates that primary treatment reduces 60 per cent suspended solids, 35 per cent BOD, 30 per cent COD (Chemical Oxygen Demand), 20 per cent nitrogen and 10 per cent phosphorus.

Secondary Treatment

Secondary treatment of waste involves the biological degradation of organic material by micro-organisms under controlled conditions.

The usual method is to bring about the biological oxidation of the organic material under aerobic conditions, in which the waste is aerated to supply oxygen for the micro-organisms. The degraded material settles out in secondary settling tanks and afterwards removed by sedimentation. The clarified waste water is discharged in the outfalls to rivers, lakes or oceans (see Figure 8.5).

Tertiary Treatment

Tertiary treatment aims at further purification of waste water and also for its recycling. The number of methods used for tertiary treatment are:

(i) chemical coagulation and filtration,
(ii) adsorption,
(iii) chemical oxidation,

(iv) desalination, and
(v) oxidation ponds.

However, there is a growing need for advanced procedures that will provide a product capable of being reused for various purposes. The recycled waste water can be used for irrigation, mainly for non-food crops such as grasslands, lawns, play grounds, etc. The renovation of waste water to a quality that would permit its reuse for a variety of purposes is a major objective of current research in this field.

References

Owen, S. Oliver (1971), *Natural Resource Conservation: An Ecological Approach*, Macmillan, p. 141.

Manivasakam, N. (1992), *Environmental Pollution*, NBT, p. 34.

Gilpil, A. (1978), *Dictionary of Environmental Terms*, p. 171.

9

Noise Pollution, Land/Soil Pollution, Pollution by Radioactivity and Their Management

Noise Pollution

Sound is a normal feature of our life and a medium through which communication is possible. It may be pleasant, like the sound of running water of a stream or of birds in the forest or that produced by musical instruments. On the other hand, it may be noisy and harsh sound of lightning and thunder or that produced by machines, automobiles, railways, aeroplanes or even the blaring sound of loudspeakers or some musical instruments. But all sounds are not noise. According to Maxwell (1973), "Noise is any sound that is not wanted. It is one of the more common forms of atmospheric pollution." Sound is caused by the vibration of molecules, while noise radiates from vibrating surfaces. More and more noise is the creation of modern civilisation and has now become a major environmental pollutant, specially in urban areas. In 1972, the UN Environment Conference at Stockholm, noise pollution has been accepted as a problem, which needs proper attention.

The problem of noise pollution has only recently received much attention, yet it has been decades in the making. The perception of noise as a problem is different for different people. Noise at certain levels might be an aggravation or nuisance for some, while at higher levels, it might become a more widespread or general aggravation. At still higher levels, noise might fall within 'discomfort range' and at even higher levels, it may lead to physical and psychological damage.

Measurement of Sound/Noise and Its Intensity

The most popular unit for measuring the noise level is the *decibel.* It is measured by an instrument known as decibel meter. A sound between 0 and 1 decibel is about the weakest that an average human can hear. For testing purposes, 0 decibel is considered to be the threshold of hearing. A whisper is about 20 decibels and an average speaking voice is about 60 decibels. The loudest sound that a person can hear without discomfort is about 80 decibels. Automobile horns may reach 90 decibels and a jet airplane may have an intensity of about 140 decibels. The levels of common noise are given in Table 9.1 and have also been depicted in Figure 9.1.

Table 9.1

Intensity of Various Sounds

Sound Source	*Intensity (in decibels)*	*Response Criteria*
	0–10	just audible
Broadcasting studio	20	—
Soft whisper (15 feet)	30	very quiet
Library, bedroom slow radio	30–50	quiet
Light auto traffic, air conditioning unit	50–70	intrusive
Freeway traffic, freight tram	70–80	annoying
Motor cycle, heavy traffic	80–100	very annoying; hearing damage
Rail noise	110–130	intolerable lighting-thunder
Carrier deck jet operation	140–150	painfully loud

In most of the countries, maximum limit fixed for noise is 75 to 85 decibels. According to scientists, noise above 90 decibels for a long time may result in deafness. In most of the cities in India, the noise level is 75 to 90 decibels but is more in industrial areas, near aerodromes, railway stations, highways, etc. But, since noise pollution is increasing day by day, planning for its control should be done. Actually, loudness alone is not the sole cause of noise

Figure 9.1
Levels of Common Noise

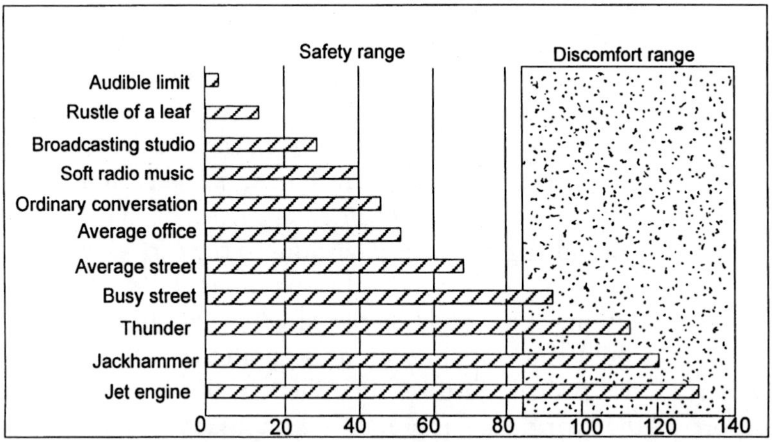

problem, although this the most apparent characteristic of noise that creates environmental problems. Pitch of the noise also influences the degree of annoyance—the higher the pitch, the greater the annoyance. Similarly, duration of the noise also determines how annoying or irritating noise will be. The degree of noise in an enclosed space or indoors is different from the level of outdoor noise.

Sources of Noise Pollution

The sources of noise can be divided into two categories, viz., natural and artificial. *Natural sources* are associated with natural phenomena like lightning, thunder, volcanic eruption, earthquake, sound of the ocean waves, etc. These reasons are local or regional and generally their impact is very limited and not very harmful. On the contrary, noise produced by man is much greater and harmful.

Artificial sources are mainly responsible for noise pollution. The sources of noise pollution in an industrial society are manifold. In cities these are associated with transportation, automobiles, motorcycles, rail engines, buses, fire brigade, police cars, ambulances, aeroplanes, industries, high pitched musical instruments, etc. During religious or domestic functions and in political meetings during elections noise pollution often increases due to the use of electronic medium of sound.

The main source of noise, however, comes from transport. The number of vehicles, particularly diesel engine vehicles, has increased the level of noise enormously. This problem is more in metropolitan cities where a large number of vehicles ply simultaneously and their noise is unbearable. The horns and sirens used by these vehicles, specially the pressure horns, are very noisy. The value of horns and sirens, except in emergencies, is questionable. The whistle of rail engines is very harmful to the ears. Busy traffic streets, highways, bus stands and railway stations all have problems of noise pollution. Larger and faster jet aircrafts have become one of the most troublesome sources of noise pollution. For example, O'Hare International Airport, in Chicago (USA), the busiest airport in the world, has more than a thousand operational jet flights on an average day. This means a flight in or out every 40 seconds. The residents within 20 km radius from the airport are facing great problems not only in their conversations, television viewing, sleep and relaxation but it has also started affecting their health adversely. The air traffic has increased to such an extent that nobody likes a new airfield in his neighbourhood.

Noise associated with manufacturing, building construction, mining, road building, etc., are second in intensity to transportation noises. But, their impact is felt more by factory and other workers. It is reported that noise inside factories can become a health hazard causing deafness. Industrial noise, particularly from mechanical saws and pneumatic drills, is unbearable and a nuisance to the ears. The impact of noise in a factory varies with age groups. For workers below 30 years of age, the 95 decibel noise is not harmful, but for 40 to 50 years age group this limit is 84 decibels and for 50 to 60 years age the bearable limit is 80 decibels. In mining, noise produced by blasting and digging is also a source of uncomfortable noise.

Nowadays, music has also become noisy. In a closed space, high volume music is harmful and is responsible for noise pollution. Similarly, during social and religious functions, marriages and other gatherings, including election meetings, the use of high volume loudspeakers has become a common feature and an irritating cause of noise pollution. The popular 'pop music' may be entertaining but its ultimate impact is not healthy on man.

Effects of Noise Pollution

The most notable effect of noise pollution is on hearing. Violent noise can cause temporary or permanent impairment of hearing, thus a cause of deafness. Continual noise can lead to gradual decline in auditory acuity and eventual deafness. Acute damage occurs to the ear drum when exposed to very loud sudden noises.

Noise causes several undesirable effects also. Damage to hearing and loss of sleep are only two of the obvious results. Noises produce irritability and a feeling of fatigue and may reduce a worker's efficiency. There is evidence that noise is one of the major causes of stress and anxiety. It also has its impact on blood pressure and is also a cause of other cardiovascular diseases. The rate of heart beat may be affected by noise. A recent report indicates that blood is also thickened by excessive noises. Eosinophilia, hyperglycemia, hypokalaemia and hypoglycemia are also caused by a change in blood and other body fluids due to noise. Noise also causes headaches and irritability. The effects of noise on the foetus are not fully known. Medical scientists have noted that an unborn child will move and kick when there is a loud noise. Some of studies indicate that children born in areas of high noise have some defects in comparison to the children born in peaceful areas.

Management of Noise Pollution

Since noise pollution is created by man, it can be controlled by adopting certain measures. Some of these measures could be as follows:

(i) Noise producing industries should be located away from residential areas.

(ii) Inside industries proper arrangements to minimise noise should be done by constructing soundproof walls and also to provide such instruments to workers which can protect their ears from noise.

(iii) Old machines often create more noise, therefore, all such machines should be well maintained and replaced, if possible.
(iv) The automobile horn should be designed in such a way that the noise it produces may not be harmful.
(v) Use of horn should be minimum and pressure horn should be banned as has been done in many countries.
(vi) The noise created by musical instruments, and other indoor equipments, can be checked by individuals in their own interest.
(vii) The noise created by railways can be checked by construction of ballastless rail tracks.
(viii) Special arrangements should be made to check noise near air bases. Control of aircraft noise requires several changes, which should be done.
(ix) Dense tree plantations around or nearby areas of noise can also reduce its intensity.
(x) Every government has enacted certain laws to control noise pollution. The basic need is the proper implementation of these laws and regular supervision. In countries like India, the problem of noise pollution is still in its primary stage, therefore, measures to contain it are necessary.

Land/Soil Pollution

The Problem of Solid Wastes Disposal

Although one-fourth of the surface of the earth is land surface, half of this surface is not useful to man due to permanent snow cover, deserts, mountains, etc., and only about 448 lakh sq km of land can be used by man for his multifarious activities. Any misuse or wastage of land may create problems not only for the present generation but for the future generations as well. The problem of land and soil pollution is growing day by day with the rapid growth of population, urbanisation, industrialisation, agricultural and other developmental activities. Land pollution is the result not only of man's misuse of land but more due to solid wastes disposal. The disposal of solid wastes is more a problem of developed nations but now it has also become a grave problem of all the countries of the world. Solid wastes are dumped in the ocean or open grounds; mining wastes are produced at the rate of millions of tonnes a day; slag heaps and mill tilings accumulate near processing operations; and industrial refuse contaminates streams and lakes. Only in the USA solid wastes produced annually include 30 million tonnes of paper and paper products, 4 million tonnes of plastics, 30 billion bottles, 60 billion cans, 100 million tyres, 8 million junked and abandoned automobiles and millions of major appliances of innumerable makes, sizes and kinds. No one knows the total amount of the world's wastes, but it is certain that the problem of solid wastes disposal now has become a threat to our environment, specially the environment of our cities.

Land pollution can be defined as the "changes in physical, chemical and biological conditions of the soil through man's intervention or misuse of land and resulted into degradation in quality and productivity of the soil". The problem of land pollution differs from that of water and air pollutions in the respect that the pollutants remain in place for relatively longer periods. Rapid urbanisation with the consequent increase in buildings has resulted in the reduction of lands for the wastes to be disposed of. Dumping of industrial and municipal wastes causes leaching by toxic materials which seep into the soil and affect the ground course. Agricultural practices introduce pesticides, fertilisers and manures to the land resulting in both biological and chemical contamination. Thus, land and soil pollution is increasing day by day by hazardous materials and micro-organisms enter the food chain or water and are consequently ingested by man. Therefore, there is an urgent need not only to control the soil pollution but also for its proper management.

Sources of Land/Soil Pollution

The land/soil pollution is the result of several sources, which can be categorised under following heads:

(i) domestic and municipal wastes,
(ii) industrial and mining wastes,
(iii) agricultural wastes,

(iv) radioactive materials, and
(v) biological agents.

Domestic and Municipal Wastes

One of the main causes of land and soil pollution is the growing quantity of domestic and municipal wastes. Household garbage includes food scraps, old newspapers, a variety of plastic items, bottles, discarded papers, wood, lawn trimmings, glass, canes, old appliances, tyres, worn-out furniture, broken toys and a host of other items. The total quantity of solid wastes is large and increasing. In the United States, municipal solid wastes averaged 1.2 kg per person per day in 1920. The quantity rose to 2.3 kg in 1970 and 3.6 kg in 1980 and now it is more than 4.5 kg. Urban wastes comprise both commercial and domestic wastes including dried sludge of sewage. In general, all the urban solid wastes are referred to as 'refuse'.

The amount of solid wastes generated is directly related with prosperity. In contrast to the 3.6 kg per person per day in the United States, residents of Australia produce only 0.8 kg per person per day. The average person in India produces only about 0.2 kg per day. The general composition of municipal wastes is as follows:

Table 9.2

Average Composition of Municipal Refuse in Developed Countries (% by weight)

Paper wastes	–	42
Wood and bark	–	2.4
Grass	–	4.0
Brush	–	1.5
Cuttings green	–	1.5
Leaves dry	–	5.0
Leather goods	–	0.3
Rubber	–	0.6
Plastic	–	0.7
Oil, paints	–	0.8
Linoleum	–	0.1

Contd...

Contd...

Rags	—	0.6
Street refuse	—	3.0
Household dirt	—	1.0
Unclassified	—	0.5
Total refuse	—	64
Garbage	—	10.0
Fats	—	2.0
Total food wastes	—	12
Metals	—	8.0
Glass and ceramics	—	6.0
Ashes	—	10.0
Total non-combustibles	—	24
Grand Total	—	100

It has been estimated that in 45 major cities of India, the quantity of average per day municipal wastes is about 50,000 tonnes. The National Environmental Engineering Institute (NEERI), Nagpur, has given details of the municipal wastes in Indian towns as follows:

Table 9.3

Nature of Solid Waste in Indian Towns

(in percentage)

Material Composition	*Towns According to Population Size*			
	upto 2 lakh	*2 to 5 lakh*	*5 to 20 lakh*	*above 20 lakh*
Paper	3.09	4.74	3.80	7.07
Plastic	0.57	0.59	0.81	0.86
Metals	0.51	0.39	0.64	1.03
Glass	0.29	0.34	0.44	0.76
Ash and dust	46.60	39.97	41.81	31.74
Mixed material	33.41	39.76	40.05	41.74
Carbon	12.36	12.51	11.95	15.92
Nitrogen	0.60	0.61	0.50	0.51
Phosphorus, phosphate	0.70	0.71	0.67	0.59
Potassium	0.70	0.73	0.72	0.67

It becomes clear from the above table that in our cities 90 per cent wastes include ash, dust, mixed material and carbon, while in developed countries paper, plastic, glass, metal, etc., dominates.

The dumping of domestic and municipal wastes is a serious problem in cities because of its impact on environment and public health. Solid wastes may or may not cause diseases in man but are hazardous to health. Diseases such as dysentery, diarrhoea, plague, malaria and numerous others are the result of the indiscriminate dumping of wastes.

Industrial and Mining Wastes

The disposal of industrial solid wastes is the major source of soil pollution by toxic chemicals. The industrial wastes are mainly discharged from coal and mineral mining industries, metal processing industries and engineering industries. They contain toxic metals such as lead, copper and chemicals having acids and are responsible for soil pollution. It has been reported that about 50 per cent of raw materials ultimately become waste products in industry and about 15 per cent of it are toxic. The chemicals discharged from the industries often enter the surface or groundwater or poison the soil or crops. The production of consumer goods also involved environmental problems, including land and soil pollution. The following table lists some of the effects of the production of consumer goods:

Table 9.4

Production of Consumer Goods and its Effects on Land/Soil

Process	*Activity*	*Effects*
Extraction	Mining and oil drilling	Surface land disruption, acid mine drainage, mine failings, sludge ponds, oil spills.
	Agriculture	Disruption of natural habitats, soil erosion, fertiliser run-off, poisonous pesticides.
	Forestry	Habitat disruption, soil erosion, pesticides.
Processing	Manufacturing	Soil and underground water pollution, loss of soil productivity, air and water pollution.
Energy	Energy conversion and transmission	Depletion of resources, thermal and radioactive pollution, disruption of land for transmission rights of way. Litter, land-use disruption, release of hazardous wastes, dust, etc.

The expansion of mining activities in many countries has now become a major cause of land pollution due to loss of soil and destruction of land. Mining has become a threat to the environment because it leads to the huge quantity of waste generally not useful to man and its reprocessing is not economical. Mining operations produce about 1.35 billion tonnes of debris each year.

Agricultural Wastes

Agricultural practices also pollute the soil. According to an estimate, agricultural activities produce more than 1.8 billion tonnes of waste, each year. About three-quarters of this is manure. Much of this manure is piled in dumps where it pollutes streams and waterways. Yet, at the same time, farmers across the continent are suffering from worn-out and depleted soils. Other agricultural wastes include branches and slash left over from logging apart from animal wastes. In addition to fertilisers and pesticides, soil conditioners and fumigants are used in agriculture. Organic compounds containing lead, mercury and arsenic, when applied to a land, accumulate on the soil permanently and introduce these toxic metals into plant products.

Radioactive Materials

The radioactive wastes produced by nuclear testing laboratories and industries reach the soil and accumulate there. Wastes from nuclear reactors contain ruthenium-106 and rhodium-106, iodine-131, barium-140, lanthanum-140, cerium-144, etc. All the radionuclides deposited on the soil emit gamma radiations, and are harmful to soil as well as for plant growth.

Biological Agents

The excreta of humans, animals and birds is also a source of soil pollution by biological agents. Digested sewage sludge, which is used as a manure, also causes soil pollution. In the developing countries, intestinal parasites constitute the most serious soil pollution problems. Faulty sanitation, waste water and wrong methods of agriculture also induce soil pollution. There are three groups of pathogenic organisms that pollute the soil, viz., (i) pathogenic organisms excreted by man, (ii) pathogenic organism excreted by animals, and (iii) pathogenic organisms found in contaminated soil.

Harmful Effects of Land/Soil Pollution

The following are the harmful effects of the land/soil pollution:

(i) Land/soil pollution is responsible for loss of fertility and productivity of soil.
(ii) The decomposition of the various types of waste material causes harmful gases and bad smell, which not only pollute environment but is also harmful if mixed with chemicals.
(iii) The municipal and domestic waste is often discharged in water bodies, thus responsible for water pollution.
(iv) The dumping of wastes in oceans is a common practice and a main cause of imbalance of the marine ecosystem.
(v) Particles in the sewage may clog the micro holes of the soil and also destroy the micro-organisms necessary for the soil.
(vi) Industrial and chemical wastes are also responsible for the pollution of underground water.
(vii) Those bacteria which are transmitted from man to soil infect man causing dysentery, cholera, tuberculosis, typhoid, paratyphoid fever, etc.

In fact, land/soil pollution is one of the main causes of air and water pollution, therefore, its control and proper management is necessary not only in developed countries but also in developing countries.

Management of Land/Soil Pollution: Methods of the Disposal of Wastes

Land/soil pollution can be controlled by adopting measures for the disposal of solid wastes, which unfortunately involves large finances and technology. Prior to disposal, the wastes have to be collected from the places by trucks to the disposal area. It is estimated that about 80 per cent of the total cost of solid waste management is spent on collection only. The methods used for the disposal of waste are open dumps, ocean dumping and sanitary landfills.

Dumping is a popular and an inexpensive means of solid wastes disposal. *Open dump* is the least acceptable method but prevalent in most of the developing countries, including India, and is a cause of soil as well as general environmental pollution. In this method, waste is collected and hauled to a site where it is dumped on the ground. Organic matter rots or is consumed by insects, rats or hogs. In some communities, the pile is set a-fire in the evening to reduce the total volume and the odour. All this creates serious ecological problems. The dump itself is a potential source of disease. Rain erodes the dump, and the polluted water flows into nearby rivers and groundwater reserves.

Sanitary landfill is commonly used for final disposal of solid wastes. This is cheaper than other disposal methods and have the advantage of avoiding acute pollution problems associated with discharging wastes into waterways or polluting air from incineration. But, the main disadvantage over open dumps is in the aspect of public health. Leaching by surface water or groundwater may contaminate streams and other water supplies. The availability of land for the dumping of the growing volume of wastes is also a problem.

Ocean dumping of wastes is practised by many coastal cities. It was considered to be the favourite disposal method under the prevalent notion that the ocean is an inexhaustible sink. We are now learning that the ocean does not have an infinite capacity for covering, holding, absorbing or decomposing materials, nor the capacity for anywhere near the load of pollution that would result if the rate of ocean disposal continues, as in the past. There were seven major categories of waste—harbour dredgings, industrial wastes, municipal sewage sludge, refuse and garbage, construction and demolition debris, military explosives and chemicals and miscellaneous wastes like agricultural land run-off, pesticides, fertilisers, sewage and garbage from vessels, oil spills, etc. Now, the environmentalists have warned that ocean dumping is neither a safe nor a desirable method of the disposal of wastes. In 1970, the Council on Environmental Quality recommended a series of drastic actions to control ocean dumping because it upsets the ecological balance of regions of the sea.

Incineration is a method through which the volume of wastes can be reduced by 60 per cent and it also reduces the public health problems. This method is adaptable over a wide range of capacities from small domestic incinerators to large centralised municipal plants. It can handle a mixture of garbage and rubbish, and the physical nature of the residue—called *clinker*—aids in the disposal of the waste, although it aggravates the problem of air pollution.

Chemical processing of solid wastes is also applied but it is costly and technically not feasible for poor countries. But, due to prospect of energy production and recovery of certain usable materials, the salvage of some wastes by chemical methods is now considered economically feasible. The reclamation of non-ferrous scrap metals has long been an established industry.

Composting municipal refuse to convert it into a fertiliser and soil conditioner is appealing not only because it appears to be a good way to recycle the resources in solid wastes, but also because of the beneficial nature of the compost. This may be considered as useful method for the disposal of municipal refuse, specially in developing countries.

New methods of waste disposal that are under trial in various parts of the world are:

— The transformation of the organic content into sugar or proteins.
— The heating of organic refuse under anaerobic conditions to convert it into useful gases such as methane, which is usable as fuel, or into a liquid product.
— The compression of refuse into building blocks that can be sheathed with more durable materials. One such plant is in operation in Japan.
— The compression of refuse into briquettes that can be used as fill. A mixture of fly ash, dried sewage sludge, incinerator residue and river and lake dredgings is under study.
— The use of degradable plastic bags instead to trash cans in the collection and transport of municipal refuse.
— The transport of refuse as a liquid slurry in pipelines, a method now under study.

Energy from Refuse

In many metropolitan areas of developed countries the garbage is burnt in a well-engineered furnace. The heat from the fire is used to boil water and produce steam. Then, the steam is sold for industrial use. But, there are problems in this type of burning also as wet garbage and food scraps are difficult to burn and burying PVC products produces hydrogen chloride gas, which reacts with water to produce a strongly corrosive liquid—hydrochloric acid. In spite of these difficulties, incineration may become profitable in the future because:

- Increasingly large quantities of dry paper and cardboard have appeared in refuse, thereby increasing the fuel content of the trash.
- The price of fuel has skyrocketed and therefore the value of steam has also gone up, thus it will be more economical.
- The rising cost of land has made it harder to find adequate sites for landfills, thus demanding alternative solutions and burying of refuse is the most suitable alternative.

Recycling of Waste Material

Most refuse contains a wealth of raw materials that can be easily reused or recycled. A large quantity of metals can be obtained from waste dumps, as also paper, glass, tyres, manure, etc. Table 9.5 indicates the possibilities of recycling the various waste materials.

It becomes clear from the table that if the recycling process is adopted properly, not only the problem of waste disposal can be minimised but it will save financial resources also. An estimate indicates that about 21 per cent of the paper in United States was recycled, compared with about 50 per cent in Japan. If the rate of recycling were increased to 50 per cent in United States, approximately 100 million trees would be saved annually, and the energy thus saved would be enough to supply 7,50,000 homes with electricity. If people would separate their household trash into piles of aluminum, steel, paper, plastic and food scraps, then, perhaps, all these materials could be recycled.

Table 9.5

Various Recycling Routes for Some Common Wastes

Waste	*Recycling Possibilities*
Paper	Use the backs of business letters for scrap paper or personal stationery, lend magazines and newspapers to friends
	Repulp to reclaim fibre
	Compost
	Incinerate heat
Glass	Purchase drinks in deposit bottles and return them, use other bottles as storage bins in the home
	Crush and remelt for glass manufacture
	Crush and use as aggregate for building material or antiskid additive for road surface
Tyre	Recap usable casings
	Use of swings, crash guards, boat bumpers, etc.
	Grind and use as additive in road construction
Manure	Compost or spread directly on fields
	Ferment to yield methane; use residue as compost
	Convert to oil by chemical treatment
	Treat chemically and re-use as animal feed
Food scraps	Save for meals of leftovers
	Sterilise and use as hog food
	Compost
	Use as culture for yeast for food production
Slaughter house and butcher shop-wastes	Sterilise and use as animal feed
	Render
	Compost

New Approach

The above mentioned methods have proved quite useful for solid waste management. But, now-a-days, there is more emphasis on people's participation. There is a growing adoption of the polluter pays principle, i.e., NIMBY and NIABY.

The Polluter Pays Principle: NIMBY and NIABY

The public are becoming better informed on environmental issues, and may object to real or imagined pollution and waste hazard or nuisance. Special interest groups or the public can adopt a "not-in-my-backyard" (NIMBY) attitude. NIMBY is often invoked when authorities are proposing to site a hazardous or nuisance waste or pollution treatment plant. Some proposals are opposed by environmental activists and NGOs whenever they are to be sited (for example, deep underground storage of high-level nuclear waste)—a "not-in-anybody's-backyard" (NIABY) response.

Apart from the polluter pays principle, over last few decades, pollution and waste management have been aided by:

— a trend towards more proactive planning and management;
— increased release of information on pollution and waste risks, as a result of government, NGO, media and international agency activity;
— improved legislation;
— development of better policing;
— spread of better environmental quality standards more widely applied;
— better methods and equipment for monitoring and assessment; etc.

Radioactive Pollution

Nature and Sources

Radioactive substances are among the most toxic materials known. The magnitude of their injurious effects is in a range that might be said figuratively to represent a 'quantum jump'. H.P. Jammet, Chief of Atomic Hygiene and Radio Pathology, French Atomic Energy Commission, defines radioactive pollution of the environment "as the increase in natural background radiation, emerging from the activities of man involving the use of naturally occurring or artificially produced radioactive materials. The atmosphere screens out much of the sun's radiation, including most of that which would be lethal to life".

Naturally occurring, radioactive elements are present in the rocks, water and air, and in all living organisms. The composite of all forms of natural radiation that we are exposed to, is called background radiation. For millions of years, background radiation remained at a relatively constant level. Recently, however, the radiation that we are normally exposed to has increased and is expected to increase further. Man-made radiation includes that from X-ray machines, radioactive fallout, nuclear reactors, research laboratories, industrial and medical uses of radioactive materials and radioactive wastes from the production of radioactive materials.

Environmental radiation may be divided into two types: (i) natural, and (ii) man-made. Naturally occurring radiations are: (i) cosmic radiations from the outer space reaching the earth's surface, and (ii) terrestrial radiation from natural radio-isotopes present in the earth's crust. Cosmic ray bombardment of the atmosphere continuously produces a few radioactive materials of short half-life. The half-life of a radionuclide refers to its period of radioactivity and has a bearing on the length of time that a radioactive substance might remain active as an environmental pollutant. The following table indicates the half-life and radiation of the important radioactive elements:

Table 9.6
Commonly Used Radionuclides

Material	*Half-life*	*Radiation*
Carbon-14	5, 568 yrs	beta
Cobalt-60	5.24 yrs	beta, gamma
Hydrogen-3(tritium)	12.26 yrs	beta
Iodine-131	8.08 days	beta, gamma
Phosphorus-32	14.3 days	beta
Plutonium-239	24, 360 yrs	alpha, gamma
Radium-226	1622 yrs	alpha, beta, gamma
Strontium-90	27.7 yrs	beta
Uranium-235	710 million yrs	alpha, gamma
Uranium-238	4.51 billion yrs	alpha, beta, gamma

Source: Maxwell, K.E., *Environment of Life*, 1973, p. 230.

Man-made radiations originate from the activities of man involving the use of radioactive materials. They are used for the production of nuclear weapons, nuclear fuel and electric power.

One of the most harmful uses of radioactive material is in nuclear weapons, specially in atom bombs. The world has already seen the nature of destruction as well as impact of radioactivity on man and environment during the Second World War when atom bombs were dropped over Hiroshima and Nagasaki—the two Japanese towns. During the last forty years, there have been a number of nuclear weapon testings either in sea or underground, resulting in the increase in the level of radiation of the environment. A study revealed that the radioactive materials which cause such explosions are fission products as Strontium-90, Caesium-137, Iodine-131 and unused explosives. These elements either settle down in the soil or water or remain suspended in the atmosphere, thus becoming a major cause of radioactive pollution.

Nuclear fuels used in the operation of reactors also contribute to radioactive pollution. The heat liberated during fission and activation process is converted into electricity. During this process, two types of wastes are formed: (i) the fission products remaining in both the primary and secondary fuels, and (ii) extraneous activation products in the coolant. The disposal of radioactive or nuclear wastes is a great problem and a cause of environmental pollution. It includes remaining metal products at the site of mines, fission products and activation products. These are of three categories, viz., low level, intermediate level and high level. Among these, high level products remain in the environment for several hundred years. Similarly, leakage in nuclear reactors often becomes the cause not only of environmental pollution but also of the death of hundreds of people as happened in case of leakage at Chernobyl Atomic Reactor in erstwhile USSR on 26 April 1986.

Effects of Radioactive Pollution

Different forms of radiation have different effects. Visible light and infrared heat rays and certain kinds of radiation are generally beneficial. Some of the other forms of radiation, for example, X-rays, are more energetic and are, therefore, biologically injurious because of their destructive action on cells and tissues. Because these highly energetic forms of radiation tend to split substances, including living matter, into ions, they are called ionising radiation.

The other forms include electromagnetic radiation and particulate radiation. Ultraviolet radiation is highly injurious to the tissues of some organisms. It is well recognised that sun-burn is caused largely by the ultraviolet content of the sun's radiation and that excessive exposure can cause serious injury to both the superficial and deep layers of the skin.

The environmental hazards from fission reactors include: (i) the possibility of a serious accident that releases a large amount of radioactivity, (ii) the hazard of even the usual small radioactive emissions with normal operations, and (iii) the pollution caused by radioactive wastes. The fate of a radioactive contaminant is also dependent on ecological conditions. Radioactive pollution causes deposition of the contaminants on ground and surface waters. The water consumed by plants thus acts as a medium for radioactivity. This contamination is passed on to vegetation and animals through the nourishment from soil and water.

Man is considered to be the ultimate sufferer through reactions and interactions. Radioactive particles present in air have a direct impact on man's respiratory system, while indirect contamination occurs by consumption through the food chain. The ill-effects of radioactive pollution are numerous. Radiation can cause cancers, abnormal births, skin diseases and mutations in man. Chronic exposure to radiations leads to leukemia in an individual and even affects an unborn child in the womb. The radiations from medical X-rays, colour TV, luminous dials of clocks and watches and X-ray fluoroscopes are also contributing to the risks. As described earlier, any leakage of radioactivity from reactors is not only a health hazard but also a cause of death and other disorders to thousands of people. Similarly, nuclear testing in sea waters not only creates disorder in the oceanic ecosystem but becomes a cause of the destruction of marine life too.

Management of Radioactive Pollution

In spite of certain harmful effects of radioactivity, nuclear power and other uses of radioactive elements have become an integral part of modern development. Many of the environmentally conscious countries in the world depend on nuclear power significantly for their electricity generation. At present, 17 per cent of the world's electricity is generated through nuclear sources. Nuclear source is considered clean, compact and concentrated. It is clean because its impact on the environment is minimal compared to other energy generating sources. To make it environmental safer still, there is a need for monitoring and prevention measures.

Monitoring of radiation is very necessary for its control. For this, sampling and its proper analysis are necessary. The samples of air, water and soil need to be collected regularly and their analysis—both manual and instrumental—be done, not only to know the actual level of radioactive pollution but also for its control.

Prevention and control measures are to be adopted to control radioactive pollutants. All measures should so aim that the level of radioactive pollution should not exceed the maximum permissible limit. The industrial wastes having radioactive elements only should be discharged after proper treatment. The high activity wastes should not be discharged. Now, a few techniques have been developed through which these wastes can be used, if treated properly. The disposal of the wastes into sea is also not a safe method but if possible, these wastes should be sealed in concrete filled steel drums and discharged into depths of 1,000 fathoms or more.

The emission of radioactive pollutants should be controlled at any cost. In uranium mines, wet drilling can be employed. Closed-cycle systems with gaseous coolants of very high purity may be used in the case of nuclear reactors to prevent extraneous activation products. There is also a need to control radiation hazards for workers and officials working in reactors or other places having danger of radioactivity. A kit to protect from radiation effects and health care should be provided to all workers. What is more, workers at power stations must be trained in radiation protection and radiation hazard control measures. The UN and other agencies of the world are in favour of total ban on nuclear weapons but very little progress has been made in this field. Let us hope that the time will soon come when our world will be free from the danger of atom bombs and other nuclear weapons.

Reference

Maxwell, K.E., *Environment of Life*, Dickenson Publishing Company, California, 1973, p. 175.

10

Sustainable Development

Man has continually, and at an increasing rate, been changing the forms and modes of his interaction with the environment. Natural processes and factors that depend on the structure and other characters of the earth, and social processes and factors that depend on structure and characteristics of our society, have been closely interwoven. On the one hand, population of the earth is growing rapidly and by the year 2010 AD, it might exceed the 7.5 billion mark, while on the other hand, with the technological and scientific revolution, developmental activities have been accelerated to such an extent that in certain areas they have become a threat to the environment. The consumption of natural resources is increasing at an alarming rate, causing great concern for their conservation. The degradation of the environment, due to industrial and other wastes discharged into the atmosphere and hydrosphere, has also caused great concern and there is a big question—what will happen next? All these environmental problems associated with development have raised several questions regarding the type and nature of development and this has given rise to the concept of 'sustainable development'.

Concept of Sustainable Development

The term 'sustainable development' was first used in the Cocoyoc Declaration on Environment and Development in the early 1970s. Since then, it has become a buzzword for international organisations dedicated to achieving environmentally benign or beneficial development. The term has served to catalyse debate over the relationship between economic change and the natural resource base in which it is grounded. Definitions vary, but the most widely adopted one is contained in *Our Common Future*—the 1987 report of the World Commission on Environment and Development: "Sustainable development is the development that meets the needs of the present without compromising the ability of future generations to meet their own needs." This definition is far from adequate—it begs the critical question of what constitutes an acceptable 'need' in the context of present global situation—but it does encapsulate the essential point—that we all have to learn to live within our planetary means. Various UN publications, in an effort to define strategy for sustainable living and development, refer to it as "improving the quality of human life while living within the carrying capacity of the supporting ecosystem". Some definitions of sustainable development are:

— A strategy that meets the needs of the present generation without compromising the ability of future generations to achieve their own requirements.
— Improving the quality of human life while living within the carrying capacity of supporting ecosystems.
— Environmental care 'married' to development.
— An environmental 'handrail' to guide development.

The concept of sustainability, has a long historical background. In ancient Indian writings, 'nature' or environment has been considered as the controller of all human activities, including economic development. They have given the status of God to all the components of environment, such as air, water, land, natural vegetation, animals, etc. In spite of the fact that there was no need for their conservation, the concept of their protection has been put forward in all the religions of India. In Greek writings, there is a reference of '*Ge*' or '*Gaia*' as the Goddess of the Earth—the mother figure of natural replenishment. In fact, that was a time when population was limited and there was mutual adjustment between man and environment.

It was only after the industrial revolution, followed by transport revolution and urbanisation, that degradation of the environment started and which has given rise to the concept sustainable development. The sudden interest in sustainable development is, of course, a response to warnings that the world is facing an environmental and, therefore, social catastrophe in the near future unless mankind radically modifies certain practices and perspectives which have created the present crisis. This crisis is characterised, among other things, by the poisoning of our rivers, seas and underground water sources; the thinning of the ozone layer; global warming; the rapid extinction of species, massive deforestation and soil improvement; rapid population growth; and uncontrolled urbanisation, with its attendant social problems.

The former US President Roosevelt's chief forester and mentor, Gifford Pinchot (1910) had observed that "conservation advocates the use of foresight, prudence, thrift and intelligence in dealing with public matters, demands the application of common sense to the common problems for common good". Conservation of resources, in other words, is the 'sustainable utilisation' of the resources, which can be achieved through environmental regulation, management, adequate surveillance and accountability. The environmental degradation and associated problems, like desertification, soil erosion, tropical forest depletion, etc., in the Third World, are the consequences of present development policies. Therefore, there is need for positive mutual adjustment between development and natural environment.

Several strategies, policies and concepts have been propounded by ecologists and planners to deal with environment/natural resources management, such as the zero growth strategy of Daly (1977), organic agriculture for self-sustaining societies of Naess (1977), a narrow and overspecialised approach of environmental economics by Norgaard (1985) and the well-known concept, i.e., use of environment-friendly techniques and products.

The publication of US *Global 2000 Report* (1980) and other publications like *Resourceful Earth* (Simon and Kohn, 1984), *The Global Possible* (Repetto, 1985) and *Our Common Future* (World Commission on Environment and Development, 1987) have generated a debate highlighting the consequences of overexploitation of natural resources and neglect of global 'common interest'. The various programmes of the UNO, specially the programmes of UNEP, have emphasised the need for sustainable development, also referred to as 'eco-development'.

Despite its simplicity, the phrase 'sustainable development' is capable of carrying a wide range of meanings, sometimes allowing quite different ideas about what should be done to exist side by side without the conflict between them being clear. Both radical environmentalists and conventional development policy pragmatists have seized the phrase and used it to express and explain their ideas about development and environment. In the process they have created a theoretical maze of great complexity (LéLé, 1991; Redclift, 1996; Redclift and Benton, 1994; Adams, 2001). The superficial conformity of writing and thinking about sustainable development hides very real and fierce battles behind the scenes over the meaning.

Sustainable Development at Rio, 1992

The United Nations Conference on Environment and Development (UNCED) at Rio in 1992 (the Earth Summit) was attended by 128 heads of states and in total by the representatives of some 178 government. Debate at the conference drew very directly on the mainstream ideas about the environment and development that had evolved during the 1980s. The Rio Declaration and the much larger *Agenda 21* were the fruit of endless negotiation at a series of Preparatory Commission meetings and at Rio itself, between teams of diplomats determined to surrender as little as possible of their national interest. A key feature of these debates was the gap between countries in the industrialised North and the underdeveloped South, which became steadily more glaring in the run up to, and during, the conference. The issues of climatic change and biodiversity that dominated the Rio Conference are vitally important to certain countries (especially those vulnerable to sea-level rise), but they are not the principal environmental problems faced by most countries of the South.

The Rio Declaration ended up as a bland list of 27 principles. It was long-winded, and in places self-contradictory: even after long debate, the US delegation released an 'interpretative statement' that effectively dissociated it from a number of the principles agreed. It dissented from principle 3 that there was a right to development, the Americans arguing that development was not a right but simply a shared goal, and from any interpretation of principle 7 that suggested that there was an international liability to make development sustainable (i.e., for rich countries to pay for it). Choices about development and environment were matters for individual countries to make their minds up about, and not an issue over which they should be subject to international opinion.

Agenda 21 was even more burdened by divergent opinions, becoming a bloated volume of more than 600 pages in 40 separate chapters. Its scope was enormous, covering issues from biodiversity and water quality to the role of women, children and organised labour in delivering sustainable development. It reflected previous mainstream thinking about sustainable development in several ways. First, it made the need for economic growth a central theme, as in the Brundtland Report. In the sustainable development mainstream, everything is predicated on economic growth, both globally and nationally. Second, *Agenda 21* emphasised the familiar straightforward issues of environmental management: all the familiar environmental issues from the *World Conservation Strategy* appear developed but unmistakable. Third, *Agenda 21* was technocentric. The first six key themes make this quite clear: growth with power and technology will direct the evolution of policy towards more efficient use of the environment and hence towards a more sustainable world economy. The 'essential means' to achieve sustainability also reflect this technocentrism, building on information, science and environmentally-sound technology. Fourth, *Agenda 21* presumed a multilateral benefit to be derived from a sustainable development strategy, as did the Brundtland Report. It suggested that change would arise from the mutual interest of industrialised and non-industrialised countries, and from the concern of present generations about the future. It suggested that this shared interest would cause international financial resources and technology to flow, directed and promoted by international agencies and structured and regulated by international legal instruments. Fifth, like its predecessors, *Agenda 21* called for sustainable development through participation. As in *Caring for the Earth,* women, children, young people, indigenous people, trade unionists, businessmen, industrialists, farmers, local authorities and scientists are all summoned to play a role—a rainbow coalition to put flesh on the endless skeleton on the text of *Agenda 21.*

After Rio

At the close of Earth Summit+5—the nineteenth special session of the United Nations General Assembly, held from 23 to 28 June 1997 in New York—delegates from over 165 countries adopted the *Programme for the Further Implementation of Agenda 21.* In the final document, governments acknowledged that the global environment has continued to deteriorate since 1992, with rising levels of greenhouse gas emissions, toxic pollution and solid waste. Renewable resources, notably fresh water, forests, topsoil and marine fish stocks, continue to be used at rates that are clearly unsustainable.

To address these concerns, governments took action on several fronts at the special session, as reflected in the final document. Among other decisions, they agreed to reconfirm the political commitment to sustainable development from all members of the international community, as well as from all major groups of civil society:

- reconfirm the financial commitments;
- establish an Inter-governmental Forum on Forests under the Commission on Sustainable Development to continue policy dialogue on this issue, including more focused consideration of elements for a possible legal instrument;
- open high-level inter-governmental dialogues on fresh water, and on energy and transport, which will be taken up by the Commission on Sustainable Development at upcoming sessions;
- make a stronger commitment at the global level to such issues as tourism, changing production and consumption patterns, and eco-efficiency; and
- set a more focused work programme for the Commission on Sustainable Development.

After a decade of Rio Conference a summit conference was organised by the UN at Johannesburg (South Africa) on Sustainable Development in 2002. The main purpose of the summit was to make an assessment of the achievements after a decade of the Rio Summit. The Rio Summit had raised the problem of the prevention of deforestation, clean air and drinking water and reduce the level of poverty in developing countries. This will lead to sustainable urban development. Unfortunately, the situation since Rio has further deteriorated. Further commitment was asserted and efforts were made for mutual understanding between developed and developing nations on sustainable development, but still a long way to go.

Eco-development

The concept of eco-development implies that all developmental activities should be performed or designed in such a way that regional ecological balance can be maintained. This was originally advocated by UNEP for environmental planning and was defined as "development at regional and local levels consistent with the potentials of the area involved, with attention given to the adequate and rational use of the natural resources, and to application of technological styles" (UNEP, 1975).

Dasmam (1985) described eco-development as a triangle, one side of which is *basic needs*, the second *self-reliance,* and at the base *ecological sustainability.* Each of these variables needs to be considered, since development will not be sustainable unless poor people are involved in meeting their aspirations. In fact, "the environment and development are means, not ends in themselves. The environment and development are for the people, not people for environment and development" (Chamber, 1986). There is thus need for 'environment thinking' with 'development thinking' and 'sustainable livelihood thinking'. In Third World countries, the main consideration should be given to 'basic needs' and 'self-reliance' which can be achieved through social planning norms. While designing eco-development for a region, consideration should be given to both short-term and long-term planning. For sustainable eco-development, it is also necessary to give priority for the livelihoods of the poor; this can be achieved through political changes at the local, national and international levels. The political, economic and social forces should be directed to fulfil the needs of population in such a way that environment remains protected.

Basic Aspects of Sustainability

The question of sustainable development has emanated from overexploitation of resources as well as mismanagement of technology. The aspects, which require monitoring of sustainability, include climatic change, biodiversity, disposal of hazardous and toxic wastes, disposal of pollution-generating industries and food and ecological security.

Swaminathan (1991) has identified nine principles for success in promoting ecologically-sound agriculture. These are: (i) land, (ii) water, (iii) energy, (iv) nutrient supply, (v) genetic diversity, (vi) pest management, (vii) post-harvest system, (viii) systems approach, and (ix) location-specific research and development.

The deterioration in the ecological base in various countries in spatio-temporal terms due to irrational management of the resource and environmental systems having damaging repercussions is reflective of unsustainable policy frame and planning strategies followed so far. Its observable signs can be listed as:

(i) extensive deforestation accounting for loss of flora, fauna and some rare species;
(ii) drying up of drinking water resources and fall in the underground water levels;
(iii) intensifying rate and frequency of flood and droughts;
(iv) land degradation due to desertification, wastelands, salinity and waterlogging;
(v) deterioration in quality of air and water;
(vi) pressure of population resulting in unemployment and mass migrations; and
(vii) unplanned urbanisation and unprecedented growth of urban slums, etc.

The environmental problems are multidimensional and varied in nature in both developed and developing countries. On the one hand, there are global problems which have had their impact throughout the world. On the other hand, every country has its own development as well as environmental problems. Apart from this, regional and local problems need immediate attention. The problems created by technology transfer from developed countries to Third World countries have become a cause for concern, because in the absence of proper management it has resulted in environmental degradation. Such alarms are embodied in the unsustainable activities which may include:

(i) intensive cultivation of land without taking adequate care of soil fertility;
(ii) development of irrigation facilities without proper water management, which leads to waterlogging, alkaline or saline soil;
(iii) improper use of pesticides, fungicides, herbicides, etc., which causes soil damage and biological imbalance;
(iv) excessive trapping of underground water accounts for steep fall in underground water level;
(v) replacement of high yielding hybrid varieties leads to spread of diseases capable of wiping out the entire crop as it happened with Irish potato crop in 1985 and Bengal rice famines in 1942;
(vi) excessive use of non-degradable material like plastic creates problems of waste management;
(vii) discharge of industrial and municipal waste in water bodies leads to problems like unmanageable water pollution; and
(viii) automobiles and industries have become a major cause of air pollution.

In fact, development without proper management has become a cause of eco-destruction, for which sustainable development is the only solution. In a report for a sustainable world, following steps have been suggested:

1. A production efficiency era of minimum environmental damage costs through *energy* transition.
2. For a stable world population, a *demographic* transition.
3. From non-renewable to renewable *resource* transition.
4. By a global mutually agreed objective between South and North, *political* transition.

While all these initiatives are necessary, the critical shift required is in the perception of man. For 'industrial man' has grown accustomed to thinking that (a) the world's resources are infinite, and (b) they exist primarily to be exploited for the benefit of mankind. These assumptions are leading us to disaster. What is required is that we replace this outmoded perspective with one that sees us all as crew members of spaceship earth and fragile life-support system spinning through space.

Guiding Principles of Sustainability

The guiding principles of sustainability cut across ecological, economic, social and cultural dimensions and there are obvious trade-offs. What needs to be addressed is sustainable development—for whom, in what context, and with what objectives?

The guiding principles of ecological sustainable resource management (Hare et al., 1990) are as under:

- Inter-generation equity providing for today while retaining resources and options for tomorrow.
- Conservation of cultural and biological diversity and ecological integrity.
- Constant natural capital and 'sustainable income'.
- Anticipatory and precautionary policy approach to resource use, erring on the side of caution.
- Use of resources in a manner that contributes to equity and social justice while avoiding social disruptions.
- Limits on natural resource use within the capacity of the environment to supply renewable resources and assimilate wastes.
- Qualitative rather than quantitative development of human well-being.
- Pricing of environmental values and natural resources to cover full environmental and social costs.
- Global, rather than regional or national, perspective of environmental issues.
- Efficiency of resource use by all societies.

- Strong community participation in policy and practice, during the process of transition to an ecologically sustainable society.
- The importance of individual, local and institutional sustainability to agricultural and environmental sustainability needs to be stressed.
- Poorly conceived globalisation/trade liberalisation threatens the sustainability and existence, economically and practically, of many developing countries (Dragun and Tisdell, 1999).

Temporal dimension of sustainability becomes obvious when we try to relate it to attributes that characterise ecosystems. A miniature ecosystem of micro-organisms created in a culture flask may have a brief turnover time of less than an hour. On the other hand, a village pond, a grassland or a forest ecosystem has a turnover time that may have to be viewed in terms of decades or hundreds of years. And, the time frame becomes even greater for a landscape of interacting ecosystem.

Measuring Sustainability

It is well recognised that conservation and sustainable development represent two sides to the same coin, closely interlinked with one another, as one cannot be achieved at the expense of the other. From a human angle, such an integrated approach demands satisfying basic human needs in an equitable manner, maintenance and promotion of social, cultural and biological diversity, and indeed the ecological integrity of the system as a whole. It is in this context that one needs to view the whole question of evaluating sustainable development.

Indicators for Evaluation

Sustainability cannot be measured in a single indicator. A number of different indicators—ecological, economic and social—need to be reconciled in any evaluation of sustainable development of a landscape unit. The economic indicator is the most obvious one. Energy efficiency is an ecological indicator, which is also helpful in the decision-making process.

How does one monitor sustainable development? We need a variety of indicators. The indicators of sustainable development are varied; therefore, compromises are called for. Monitoring and evaluation have to be done using a number of diverse indicators that may be ecological, economic, social and even cultural. Basically, these indicators add up the 'value system' that a given society desires to have.

Ecological indicators may be land use and land cover changes, biomass quality and quantity, water quality and quantity, and soil fertility. Energy management becomes the key measure for achieving the best in terms of the desired objectives.

Land Use and Cover Change

By sequential monitoring of land use and land cover changes, using a variety of technologies such as remote sensing, aerial photography and ground truth measurement, one can get a reasonable estimate of land quality. Any conclusions that suggest land degradation in the long run, obviously are not sustainable. Thus, for land use decisions, on an ecological plane, a plot level analysis, leading through ecosystem level, and eventually getting to the land-scale level analysis, becomes most obvious; a similar set of units, at social scale, would be a family, a village and a cluster of villages in a given ecological landscape. This would ensure that the end-product is sustainable livelihood as a short-term strategy, leading to sustainable development of the region as part of a long-term strategy, through a participatory and interactive analytical framework.

Biomass Quality and Quantity

The quality of biomass generated on the land (biodiversity), and the quantity of biomass generated through rehabilitation measures of a given ecosystem, are important indicators. We need quality and quantity of biomass for multiple use, for ecosystem integrity as well as for sustainable use by the humans.

Water Quality and Quantity

Any land use and cover change, and indeed any land-based human activity, if unsustainable, may get reflected in groundwater recharge and the water quality. Monitoring this through sequential recharge characteristics of the systems and/or monitoring water quality in rivers, ponds and lakes in the area immediately would give an idea about the sustainability of a given developmental pathway chosen for that landscape.

Soil Fertility

Sustainable management of soil fertility implies the effectiveness with which one is able to maintain soil fertility by strengthening internal biological processes within an ecosystem, both within the soil profile and above ground, and thus minimise energy subsides through inorganic fertilisers. The lesser the energy subsidy, the more sustainable the soil fertility management becomes.

Economic Indicators

Economic indicators are very much standardised in the already prevalent developmental paradigm. Monetary output/input analysis, capital savings or asset accumulation and dependency ratio are some of the measures adopted for economic evaluation. On a macro-scale, the gross national product (GNP) is a unified indicator of national developmental level.

Social Indicators

In a social sense, 'quality of life' is a composite indicator, which is difficult to monitor, and even more difficult to reconcile with the sub-units contained within. More easily measurable *social indicators* are health and hygiene, nutrition, food security, morbidity symptoms, educational status, etc. The difficult to quantify indicators such as societal empowerment, and the less tangible ones in the area of socio-cultural values are more difficult to evaluate.

The best way to approach sustainable development is an integrated approach through full understanding of linkages between the ecological and social processes. Figure 10.1 depicts indicators of sustainability of natural resources.

To cope with increased demand of the basic requirement of life and the limited supply of the natural resources, along with considerations of environmental degradation and ecological balance, we need to make adjustments and adaptations with these complex global phenomena, and move towards sustainability of the

Figure 10.1

Indicators of Sustainability of Natural Resources

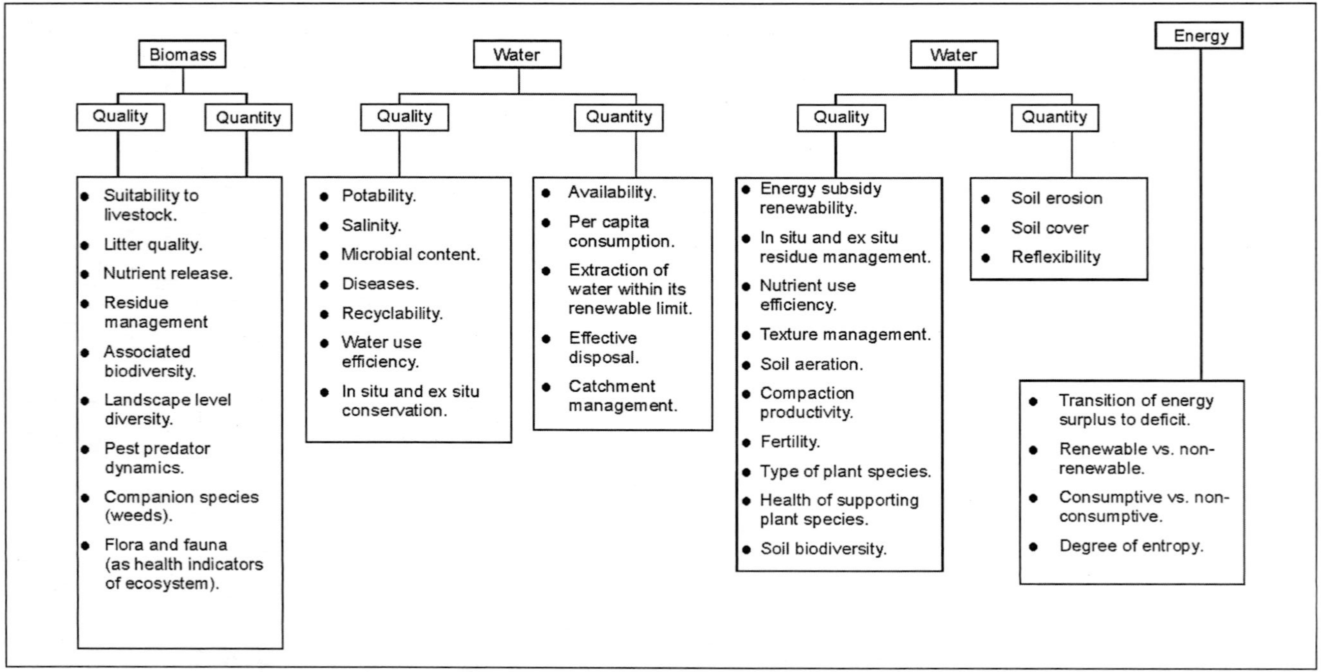

Source: Gupta, A.K., Ramakrishanan, P.S. and Malhotra, K.C., 2001, *Criteria and Indicators of Sustainability Development: A Natural Resource* Perspective, UNESCO and Oxford & IBH Pub., New Delhi, pp. 421-424.

biosphere. These were the considerations that moved an international group of ecologists from 14 countries to come together in Cuernavace, Mexico, and propose an interdisciplinary cooperative research programme—the International Sustainable Biosphere Initiative—in the early 1990s. These were the considerations that prompted the US National Research Council to set an agenda for action, which will enable the humans to complete the transitional journey towards sustainability. The council has developed an "Integrated and Location-Specific Approach for Protecting Sustainability" in the following form:

Reporting on the transition

- Indicators are essential to inform society how, and to what extent, progress is being made in navigating a transition towards sustainability.
- It needs to be understood that multiple indicators are needed to chart progress towards the goals for meeting human needs and preserving life support systems, and to evaluate the efficacy of actions taken to attain these goals.

Integrated knowledge and action

- Since the pathway to sustainability cannot be charted in advance, it will have to be navigated through trial and error and conscious experimentation. The urgent need is to design strategies and institutions that can better integrated incomplete knowledge with experimental action into programmes of adaptive management and social learning.

Priorities for research sustainability science

- Develop a research framework that integrates global and local perspectives to shape a place-based understanding of the interactions between environment and society.
- Initiate focused research programmes on a small set of understudied questions that are central to a deeper understanding of interactions between society and the environment.
- Promote better utilisation of existing tools and processes for linking knowledge to action in pursuit of a transition to sustainability.

Priorities for action: Knowledge-action collaboratives

- Accelerate current trends in fertility reduction.
- Accommodate an expected doubling or tripling of the urban system in habitable, efficient and eco-friendly manner.
- Reverse declining trends in agricultural production in Africa.
- Accelerate improvements in the use of energy and materials.
- Restore degraded ecosystems while conserving biodiversity elsewhere.

References

Barrow, C.J. (1995), Sustainable Development: Concept, Value and Practice, *The Third World Planning Review*, 17(4), pp. 369–387.

Chambers, R. (1986) *Sustainable Livelihoods*, Institute of Development Studies, University of Sussex.

Dasmam, R.F. (1985) Achieving the Sustainable use of Spices and Eco-systems, *Landscape Planning*, 12, 211219.

Dragun, A.K. and Tisdell, C. (1999) *Sustainable Agriculture and Environment: Globalization and the Impact of Trade Liberalization,* Edward Elgar, Cheltenham, U.K., p. 308.

Hare, W.L. et al. (1990) *Ecologically Sustainable Development,* Australian Conservation Foundation, Fitzory, Victoria.

Redclift, Michael (1993) Sustainable Development, *Environmental Values*, Vol. 2. pp. 3–20.

U.N. Publication, *Earth Summit+5*—Programmes for the Further Implementation of *Agenda 21.*

UNEP (1992), *Agenda-21*: Programme of Action for Sustainable Development.

11

Towards Sustainability

Sustainable development is the need of the present time not only for the survival of mankind but also for its future protection. Unlike the other great revolutions in human history—the Green Revolution and the Industrial Revolution—the 'sustainable revolution' will have to take place rapidly, consciously and on many different levels and in many different spheres, simultaneously. On the technical level, for example, it will involve the sustainable technologies based upon the use of non-renewable, fossil fuels for technologies that take advantage of renewable energies like the sun, wind and biomass, the adoption of conservation and recycling practices on a wider scale, and the transfer of cleaner and more energy efficient technologies to countries in the developing world.

On the political and economic levels, it will involve, among other things, the overhauling of development and trade practices which tend to destroy the environment, and the improvement of indigenous peoples, a fairer distribution of wealth and resources within and between nations, the charging of true cost for products which exploit or pollute the environment, and the encouragement of sustainable practices through fiscal and legal controls and incentives. On the social plane, it will involve a renewed thrust towards universal primary education and health care, with particular emphasis on the education and social liberation of women. On the environmental level, we are talking about massive afforestation projects, renewed research into and assistance for organic farming practices and biopest control, and the vigorous protection of biodiversity. On the informational level, the need is for data that will allow the development of accurate social and environmental accountancy systems.

The aim of ecologically sustainable development is to maximise human well-being or quality of life without jeopardising the life support system. The measures for sustainable development may be different in developed and developing countries according to their level of technological and economic development. But developing countries, like India, can focus attention on the following measures:

- ensure clean and hygienic living and working conditions for the people;
- sponsor research on environmental issues pertaining to the region;
- ensure safety against known and proven industrial hazards;
- find economical methods for salvaging hazardous industrial wastes;
- encourage afforestation;
- find out substitutes for proven hazardous materials based on local resources and needs instead of blindly depending on advanced nations to find solutions;
- ensuring environmental education as a part of school and college curriculum;
- encourage use of non-conventional sources of energy, specially solar energy;
- as far as possible, production of environment-friendly products should be encouraged;
- use of organic fertilisers and other biotechniques should be popularised;
- environmental management is the key for sustainable development, and it should include monitoring and accountability; and
- need for socialisation and also humanisation of all environmental issues.

The prime need for sustainable development is the conservation of natural resources. For conservation, the development policy should follow the following norms:

(i) Make all attempts not to impair the natural regenerative capacity of renewable resources and simultaneously avoid excessive pollution hampering the biospherical capacity of waste assimilation and life support system.
(ii) All technological changes and planning strategy processes, as far as physically possible, must attempt switch from non-renewable to renewable resource uses.
(iii) Formulate a phase-out policy for the use of non-renewable resources in general.

Thus, for a worldwide sustainable growth, there is need for efficient and effective management of available resources. In this field, the production of "environment-friendly products" (EFP) is a positive step. With the industrialisation and technological development, markets are flooded with products of daily consumption. They could however be a source of danger to health and damage to our environment. There is thus need to distinguish the more environmentally harmful consumer products from those which are less harmful, or have a more benign impact on the environment right from the stage of manufacture through packaging, distribution, use, disposal and reusability or recycling.

Throughout the world, emphasis is now being put on the production of EFP. In India, plans are afoot to market EFPs with combined efforts of Bureau of Indian Standards, Ministry of Environment and Forests and Central Pollution Control Board. Since 1990, a scheme of labelling *ECOMARK* has also been started. In its first phase, the items included in this are soaps, plastics, papers, cosmetics, colours, lubricating oil, pesticides, drugs and various edible items. The objectives of the scheme are: (i) to provide an incentive for manufactures and to reduce adverse environmental impact of their products, (ii) to reward genuine initiatives by companies to reduce adverse environmental impact of their products, (iii) to assist consumers to become responsible in their daily lives by providing them information to take account of environmental factors in their purchase decisions, (iv) to encourage citizens to purchase products which have less harmful environmental impact, and (v) to improve the quality of the environment and to encourage the sustainable management of resources.

Not only in consumer goods production but in the field of energy production also, environment-friendly techniques of power generation can be used. For example, in power production from coal, PFBC (Pressurised Fluidised Bed Combined Cycle) technique is useful in which coal is burnt efficiently and cleanly in combined cycle plants.

To cope with increased demand of the basic requirement of life and the limited supply of the natural resources, along with consideration of environmental degradation and ecological balance, we need to emphasise on optimal management of land, water, minerals and other natural resources. There is also need to utilize the native wisdom of those people, who live close to nature and earth, for eco-restoration along with development.

In order to apply the principle of sustainable management in reality, a highly complex way of looking at the problem is required, involving various disciplines. Sustainability is first and foremost a mental question. Without a grasp of the need or the will to change awareness, we will not succeed in realising the principle of sustainability in agriculture. It is upon the decision-makers in politics to create the right framework and the pre-conditions for a sustainable development in agriculture. Global involvement, on the other hand, must not be left out of account. Sustainability reflects our understanding of necessity and responsibility on the question for whom, for what and how production can be guided into the future in a way that is efficient, environmentally sound and sparing on resources.

Global change is an ecological phenomenon, whereas globalisation is concerned with economic change. A recent analysis of sustainable agriculture in the context of trade liberalisation and globalisation raises equally significant concern for a more informed decision-making process at local, regional and international levels. The emerging issues related to the impact of globalisation on sustainable agriculture are as follows:

- There are explicit problems with the conventional theoretical economic conditions for agricultural sustainability, especially when applied at the global level.

- The processes of trade liberalisation and globalisation will not be uniform given the ecological and institutional diversity of the nations of the world.
- There will be disparities in globalised impacts between rich and poor countries—for agriculture, industries, sustainability and environment as well as income and poverty.
- There is need for serious analysis of problems and policy initiatives, since the risk of disruption to agricultural systems and environmental deterioration, social disruption and dislocation in the poorer countries of the world is clearly very high.
- The type of production technology research, facilitated by private research, will not address the significant public good and externality issues facing developing countries.
- There is need to focus on local farming situations as a basis of dealing with global problems, especially in poor countries.
- There is need to understand local institutional situations so as to determine appropriate remedial economic policies based on institutional sustainability.
- Integrated approach is essential for research and action at the regional scale related to water, atmosphere and climate, and species and ecosystems.

The pursuit of sustainability demands choices about the distribution of costs and benefits in space and time. There is also need to take advantage of the 'traditional ecological knowledge' (TEK), which encompasses all issues related to ecology and natural resource management, both at local and regional levels. Along with political dimensions of environment-society relations, the TEK can be used for both eco-restoration and sustainable development.

Economics of Sustainable Development

The development and the environment have a two-way relationship: economic growth can have both good and bad impacts on environment. Environmental economics has sought to extend neo-classical theory, by encompassing the environment, and attaching monetary values to losses in natural capital (Redclift,

1993). The mainstream of sustainable development is based on a set of ideas that Low and Gleeson (1988) label 'market environmentalism'.

Environmental economists like Pearce (1985) have advanced thinking with economics in a number of important ways, such as:

(a) They recognise that changes in natural capital stocks involve both costs and benefits.
(b) There is also a recognition that both 'non-use' and 'use' values are important. Environmental values include 'existence values' which can be arrived at notionally, and a (theoretical) price attached to them.
(c) Environmental economics also recognises uncertainty and irreversibility as principles in resource conservation.
(d) It is also recognised that 'optimality' refers only to economic uses and efficiency, and effectively excludes and social goals for conserving resources.

Pearce (1989: 8) further declares that "we know natural capital is valuable because people are willing to pay to preserve it ... ." He also observed, "Whether there is an actual market in the asset or not is not of great relevance. We can still find out what people would pay if only there were a market."

Using economic concept, sustainability can be defined as constant stocks of both man-made and natural capital maintained over time. Economic development has always implied increasing capital stocks, but insights of environmental economics suggests the need to account for stocks of both human-made and natural separately. An alternative approach allowed the possibility of trade offs between losses to natural capital in one project and gains elsewhere.

Financing Sustainable Development

The world is facing serious environmental problems, which can be solved and controlled only when proper finances are available. The question is—who will pay? Naturally, those who are responsible for it, but at the same time, it is the responsibility of those developed countries, which are capable both technologically as well as financially. Efforts have been made in this direction but still there is long way to go.

Accepting sustainable management as an objective has been easier to achieve than that of the contentious issue of financing it. The available financial estimates vary in quality, and should be treated only as indicative and order of magnitude estimates. They do, however, provide the starting point for analysing financial requirements for sustainable estimates. This lack of robustness of the estimates also makes firm international agreements difficult to achieve.

Financing sustainable development involves following three interrelated issues:

- the level or total amount of funding needed, depending on the scope of the overall programmes and targets to be achieved in specific areas;
- terms and conditions involved in the sharing of costs between countries specifically between the developed (high income) and the developing (low and middle income) countries; and
- countrywise mobilisation of the requisite internal resources.

Funding necessary to provide sustainable development to the developing countries has been estimated in three different ways as discussed below:

Estimate 1

The two conventions of UNCED (Framework Convention of Climatic Change and Convention on Biological Diversity) established the principle that the developing countries should be refunded the full agreed incremental costs associated with measures taken under them with the Global Environment Facility (GEF) as the interim funding mechanism, However, as environmental protection and development are inextricably linked, the term 'full agreed incremental cost' has proved to be difficult to define and operationalise. The two conventions did not mention any specific cost.

The GEF was endowed with an initial funding of \$1.3 billion for a three-year period from July 1991 to June 1994. It concentrates on four key areas: global warming, biological diversity, pollution of international waters, and stratospheric ozone depletion. Since March 1994, the GEF has become a permanent financial mechanism.

The GEF is managed through a collaborative tripartite arrangement involving the World Bank, UNEP and the UNDP. In practice, the World Bank, where the facilities are located, had been the dominant partner. Up to 80 per cent of the GEF funds can be attached to the Bank's sector loans. The GEF grants up to early 1996, totalled over $1 billion to fund 113 projects in 63 developing countries, with approximately 45 per cent of the total commitments funding biodiversity projects and a further 40 per cent funding projects dealing with climatic change. A group of 30 wealthy donor countries is already committed to tripling the GEF core fund to $4 billion. The actual GEF funding of around $1 billion is regarded as too small, especially in comparison to the official development assistance ($51.7 billion in 1993) and private financial flows ($157.7 billion in 1993) to make a significant difference. The supporters of the GEF argue that although the amount involved is small, it can act as a catalyst for global sustainability. For example, by providing only the incremental cost for financing more eco-friendly technology, the GEF can optimise its leverage from the limited resources, and have a much larger impact that its own resources would suggest.

Estimate 2

The UNCED secretariat estimated the average annual (over the 1993–2000 period) cost of implementing *Agenda 21* programmes for the developing countries. The estimates were based on incomplete supporting documents and not reviewed by the national governments. The estimated annual cost was just below $600 billion, with $130 billion (22%) provided through grant or concessional requirements. This amount is put in perspective when the GDP of all low-income and middle-income countries is noted: $5,276 billion in 1994. The $470 billion ($600–$130 billion) cost to be borne by the developing countries is 8.9 per cent of their 1994 GDP.

The $130 billion concessional element to be funded by the high-income countries is 0.65 per cent of their 1994 GDP of $20,120 billion. However, the total official long-term net resource flows to developing countries in 1993 was only $51.7 billion, or 40 per cent of the estimated $130 billion. The private net flows of 1993 to the developing countries were $157.7 billion. Thus, the importance of official flows is now much less than that of private flows.

Even if the newly industrialised countries, such as: Taiwan, South Korea and Singapore, begin making overdue contributions to net official flows of assistance to the developing countries, the requisite amount is unlikely to be realised. The fiscal stringency being faced by the high-income countries, along with the ideological shift towards the private sector as opposed to the public sector solutions, makes it unlikely that the estimated concessional resource flows would be forthcoming.

The UNCED secretariat estimates were carried out according to the chapter/programme areas of *Agenda 21.* The largest sums were for the following areas: human settlements ($215 billion), fresh water ($55 billion), human health ($51 billion), agriculture ($32 billion), deforestation ($31 billion), poverty ($30 billion), biotechnology ($20 billion), and education, public awareness and training ($15 billion). It should be stressed that *Agenda 21* takes a broader view of the environment and therefore poverty reduction, trade liberalisation, health care, shelter, etc., are included as part of the environmental management effort.

Estimate 3

The 1992 World Development Report was devoted to the development and environment. It estimated that in 2000 AD the additional investment in selected environmental programmes would amount to $75 billion or about 1.4 per cent of the GDP of developing countries for that year. The major items of additional investments are soil conservation and afforestation ($15–20 billion), reducing emissions, effluents and wastes from industry ($10–15 billion), controlling main pollutants from vehicles ($10 billion) and family planning ($7 billion). The report also recommended that such investments should be made on a self-financing basis to the extent consistent with equity and other objectives.

The above three examples of financial estimates suggest that clarity in this area is much below the desired level, in spite of its crucial importance. The methods used to arrive at these estimates not always being explicit, their robustness cannot be determined. It is not clear whether in kind services are included in the estimates, and what is the level of efficiency assumed in carrying out environmental projects.

Financial resources and mechanism play a key role in the implementation of *Agenda 21*. In general, the financing for the implementation of *Agenda 21* will come from a country's own public and private sectors. For developing countries, ODA is the main source of external funding and substantial new and additional funding for sustainable development. Renewed efforts are essential to ensure that all sources of funding contribute to economic growth, social development and environmental protection.

Agenda 21: Programme of Action for Sustainable Development

Agenda 21 is a programme of action for sustainable development worldwide. This is also known as the Rio Declaration on Environment and Development, adopted by more than 178 governments at the United Nations Conference on Environment and Development, known as the Earth Summit, held in Rio de Janeiro, Brazil, from 3 to 14 June 1992.

Agenda 21 is intended to set out an international programme of action for achieving sustainable development in the 21st century. It stands as a comprehensive blueprint for action to be taken globally by governments, United Nations Organisation, development agencies, non-governmental organisations and independent sector groups, in every area in which human activity impacts the environment. It provides a broad review of issues pertaining to sustainable development, including statements on the basis for action, objectives, recommended activities and means of implementation.

Principles

Agenda 21 is based on the following principles:

1. Human beings are at the centre of concerns for sustainable development. They are entitled to a healthy and productive life in harmony with nature.
2. States have, in accordance with Charter of the United Nations should pursue their own environmental and developmental polices and make sure that they do not damage the environment of other states or of areas beyond the limits of national jurisdiction.
3. The right to development must be fulfilled so as to equitably meet development and environmental needs of present and future generations.
4. In order to achieve sustainable development, environmental protection shall constitute an integral part of the development process and cannot be considered in isolation from it.
5. All states and all people shall cooperate in the essential task of eradicating poverty as an indispensable requirement for sustainable development.
6. The special situation and needs of developing countries, particularly the least developed and those most environmentally vulnerable, shall be given special priority.
7. States shall cooperate in a sprit of global partnership to conserve, protect and restore the health and integrity of the earth's ecosystem.
8. To achieve sustainable development and a higher quality of life for all people, states should reduce and eliminate unsustainable patterns of production and consumption and promote appropriate demographic policies.
9. States should cooperate to strengthen endogenous capacity-building for sustainable development by improving scientific understanding through exchanges of scientific and technological knowledge.
10. Environmental issues are best handled with the participation of all concerned citizens at the relevant level.
11. States shall enact effective environmental legislations.
12. States should cooperate to promote a supportive and open international economic system that would lead to economic growth and sustainable development in all countries.

13. States shall develop national law regarding liability and compensation for the victims of pollution and other environmental damages.
14. States should effectively cooperate to discourage or prevent the relocation and transfer to other states of any activities and substances that cause severe environmental degradation or are found to be harmful to human health.
15. In order to protect the environment, the precautionary approach should be widely applied by states according to their capabilities.
16. National authorities should endeavour to promote the internalisation of environmental costs and the use of economic instruments, taking into account the approach that the polluter should, in principle, bear the cost of pollution.
17. Environmental impact assessment, as a national instrument, shall be undertaken for proposed activities that are likely to have a significant adverse impact on the environment.
18. States shall immediately notify other states of any natural disasters or other emergencies that are likely to produce sudden harmful effects on the environment of those states.
19. States shall provide prior and timely notification and relevant information to potentially affected states on activities that may have a significant adverse transboundary environmental effects.
20. Women have a vital role in environmental management and development. Their full participation is therefore essential to achieve sustainable development.
21. The creativity, ideals and courage of the youth of the world should be mobilised to forge a global partnership in order to achieve sustainable development and ensure a better future for all.
22. Indigenous people and their communities and other local communities have a vital role in environmental management and development because of their native knowledge and traditional practices. States should recognise and duly support their identity, culture and interests and enable their effective participation in the achievement of sustainable development.

23. The environment and natural resources of people under oppression, domination and occupation shall be protected.
24. Warfare is inherently destructive to sustainable development. States shall, therefore, respect international law providing protection for the environment in times of armed conflicts and cooperate in its further development, as necessary.
25. Peace, development and environmental protection are interdependent and indivisible.
26. States shall resolve all their environmental disputes peacefully and by appropriate means in accordance with the Charter of the United Nations.
27. States and people shall cooperate in good faith and in a spirit of partnership in the fulfilment of the principles embodied in this Declaration and in the future development of international law in the field of sustainable development.

Structure

Agenda 21 is a report of over 500 pages, comprising 40 chapters, organised into four sections, which address the major areas of political actions as follow:

Section I: Social and Economic Development

1. International cooperation to accelerate sustainable development in developing countries and related domestic policies
2. Combating poverty
3. Changing consumption pattern
4. Demographic dynamics and sustainability
5. Protecting and promoting human health
6. Promoting sustainable human settlement development
7. Integrating environment and development in decision-making

Section II: Conservation and Management of Resources for Development

1. Protection of the atmosphere
2. Integrated approach to the planning and management of land resources
3. Combating deforestation

4. Managing fragile ecosystems: combating desertification and droughts
5. Promoting sustainable agriculture and rural development
6. Conservation of biological diversity
7. Environmentally sound management of biotechnology
8. Protection of oceans and seas
9. Protection of the quality and supply of fresh water resources
10. Environmentally sound management of toxic chemicals
11. Environmentally sound management to hazardous wastes
12. Environmentally sound management of solid wastes and sewage-related issues
13. Safe and environmentally sound management of radioactive wastes

Section III: Strengthening the Role of Major Groups

1. Preamble
2. Global action for women towards sustainable and equitable development
3. Children and youth in sustainable development
4. Recognising and strengthening the role of indigeneous people and their communities
5. Strengthening the role of non-governmental organisations
6. Local authorities' initiatives in support of *Agenda 21*
7. Strengthening the role of workers and their trade unions
8. Strengthening the role of business and industry
9. Scientific and technological community
10. Strengthening the role of farmers

Section IV: Measures of Implementation

1. Financial resources and mechanisms
2. Transfer of environmentally sound technology, cooperation and capacity building
3. Science for sustainable development
4. Promoting education, public awareness and training
5. National mechanism and international cooperation for capacity-building in developing countries
6. International institutional arrangements

7. International legal instruments and mechanism
8. Information for decision-making

The programme areas are generally structured in four heads as under:

- The basis for action
- The objectives
- The activities
- The means of implementation

In fact, *Agenda 21* is a guideline for sustainable development. A programme for the further implementation of *Agenda 21* was also adopted at the ninetieth special session of the United Nations General Assembly (Earth Summit +5), held from 23 to 28 June 1997 in New York by over 165 countries. This programme was designed to reconfirm the political commitment to sustainable development from all members of the international community, as well as from all major groups of civil society.

Environmental Issues and Sustainable Development

Man, with his technological and scientific skills, has made rapid developments in agriculture, mining, industries, transportation, forestry, land management and other areas. Unfortunately, in doing so, there has been much disruption of the functioning of natural environment. Now, people have begun to realise the side-effects of the developmental activities and have started their efforts in minimising these effects by adopting methods and technology for sustainable development. Some of the important environmental issues associated with development, and the role of ecology in possible solution of these problems, i.e., sustainable development, are as follows:

Food Production and Organic Farming

The increasing demand of food for growing population has become a serious problem in many parts of the world, specially in developing countries like India, China, Indonesia, Egypt, Brazil and many other Afro-Asian countries. Agricultural scientists have been successful in increasing the quantity and quality of food and other crops. This has been possible due to application of chemical fertilisers, high yielding seeds, crop rotation practices, weed control, pesticides as well as due to mechanisation in agriculture. This resulted in the Green Revolution beginning in the late 1960s with an attempt to increase food production in developing countries. New varieties of wheat, rice and corn were developed that responded better than the traditional varieties to fertilisers, irrigation and chemicals for pest and weed control.

But, along with the increase in food production, there came an increased dependence on energy-based expensive technology and several ecological problems have emerged due to the excessive use of chemicals either in the form of fertilisers or as pesticides. Now, it has been realised throughout the world that use of chemical fertilisers and other chemicals is harmful to soil productivity and also a cause of water and air pollution. The Green Revolution has not only disturbed the ecosystem but is also a cause of social revolution. In countries like USA, farmers are now shifting from routine use of herbicides, pesticides, fertilisers, etc., to organic farming in which no chemicals are used. The only fertilisers used are organic matter, such as manure, crushed limestone, etc. This system is similar to that of ancient Indian system which is more eco-friendly. The adoption of organic farming is a way for sustainable development. By adopting proper management practices, one can develop high yielding varieties even in this system. Along with agriculture, *aquaculture* can also be developed. The best way is to develop fish farming, which fulfils the increasing demand of foods, specially that of protein.

The fundamentals of sustainability in agricultural production systems could be summarised in five major elements as displayed below in Figure 11.1:

1. *Climate*: constraints and impacts.
2. *Energy and inputs*: energy resources, fertilisers, plant protection, ecological farming, techniques and technology.
3. *Policy and management*: political dimensions including economical, cultural and social issues, research and development as well as population control policy.

Figure 11.1
Fundamentals of Sustainability in Agriculture

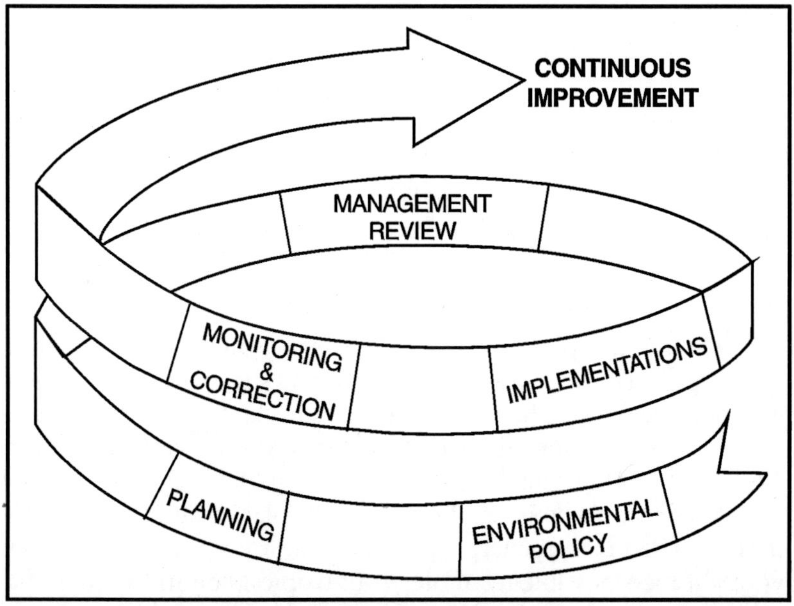

4. *Genetic resources*: identification, evaluation and utilisation of genetic resources.
5. *Soil and water*: resources and requirements.

Energy Crisis and the Ecosystem

In the ecosystem energy plays a very important role. The sun is the original source of energy on our planet. All forms of energy, i.e., nuclear, wind, coal, electric, water, gasoline, etc., originate from the sun. The total energy received by the earth's surface from the sun is enormous and much more than required by human beings. But, in modern times, the use of fossil fuels, hydroelectricity and to some extent nuclear fission are the sources of energy. The following table indicates the production and consumption of various sources of energy in the world:

Table 11.1

Production and Consumption of Energy in the World

Sources	*Production*	*Consumption*
Petroleum	126,900	127,222
Natural gas	72,791	68,290
Coal	102,322	101,660
Nuclear energy	18,373	18,373
Hydroelectricity	22,493	22,493
Total	342,872	338,038

The rapid exploitation of these energy sources has resulted in an energy crisis. In coming times fossil fuels such as oil, gas and coal will become scare. The impact of big dams and thermal as well as atomic power stations is a cause for concern to environmentalists. Therefore, for sustainable development, it is necessary to adopt non-conventional sources of energy like solar, wind, tidal, etc. Of these, the best is solar energy, provided that we are able to harness it with cheapest possible technology. In tropical countries like India, solar energy is the only alternative, which will provide an energy without any adverse impact on environment. Steps have already been taken for the development of photovoltaic cells, solar-energised water pumps, air conditioning units, solar water heaters, solar home-heating units, solar cookers, etc. Research in this field is also going on. What is needed is the development of cheaper technology to harness the solar and other non-conventional sources of energy.

Energy is directly related to the most critical social issues which affect sustainable development. These are poverty, jobs, income levels, access to social services, gender disparity, population growth, agricultural production, climate change and environmental quality, and economic and security issues. Without adequate attention to the critical importance of energy and all of these aspects, the global social, economical and environmental goals of sustainability cannot be achieved. Indeed, the magnitude of change needed is immense, fundamental and directly related to the energy produced and consumed internationally.

The importance of energy in agricultural production, food preparation and consumption is evident and essential (UNDP, 1997). The world still continues to seek energy to satisfy its needs without giving due consideration to the social, environmental, economical and security impacts of its uses. It is now clear that current approaches to energy are unsustainable. It is the responsibility of political institutions to ensure that research and development of technologies supporting sustainable systems should be transferred to the end-users. Scientists must bear the responsibility to understand the earth as an integrated whole and the impact of our actions on the global environment in order to ensure sustainability and to avoid disorder in the natural life cycle.

The key challenge in realising these targets is to overcome the lack of commitment and to develop the political will to protect people and the natural resource base. Failure to take action will lead to continuing degradation of natural resources, increasing conflicts over scarce resources and widening gaps between the rich and poor. We must act while we still have choices. Implementing sustainable energy strategies is one of the most important levers the humankind has for creating a sustainable world. Most present trends in energy indicate a deteriorating situation (UNDP, 1997). Furthermore, current energy patterns are aggravating this process by an over-preoccupation with centralised energy supply and fossil fuels to the detriment of energy efficiency, decentralised supply and renewable energy.

Pollution and Environmental Degradation

Pollution is the result of human activities associated with industrial and automobile revolutions. The forms, causes, impact and management aspects of pollution have been discussed in detail in earlier chapters. Pollution is a man-made problem, mainly of affluent countries and now a serious problem of the developing countries as well. The developed countries used to exploit the natural resources of developing countries and also export dump such materials and machines which have become a source of pollution.

The common forms of pollution are air, water, soil, noise and radioactive pollution, which can be divided into (i) non-degradable, and (ii) biodegradable pollutants from ecosystem viewpoint. The main pollutants are deposited matter, carbon monoxide, sulfur dioxide, sulfuric acid droplets, hydrogen sulfide, fluoride and other halogens, metals like lead, iron, zinc, bromine, iodine, agricultural pollutants like pesticides, insecticides, herbicides, fungicides and fertilisers, complex organic pollutants, photo-chemical pollutants from automobiles, radioactive minerals, unwanted noise, municipal wastes, etc. All these pollutants have become a threat to mankind, as they not only result in several diseases, but also environmental degradation and immediate steps should be taken to bring about a pollution-free environment. For sustainable development pollution should be checked first at source. The location of industries should be selected in such a way that their pollutants could not affect the nearby localities, and there must also be treatment plants. The various measures to check pollution have already been discussed in detail in preceding chapters. There is now need for environmental education, monitoring, environmental management and also to fix accountability and responsibility for damages, if any. The production of eco-friendly products is an innovative concept in this direction.

Problem of Waste Disposal and its Management

In developed countries enormous use of commodities in daily life is causing environmental problems. This problem is now not limited to developed countries but growing day by day in developing countries also. Wastes and toxic substances are accumulating at a rapid rate in our surroundings. These substances enter the biogeochemical cycles. In man-made ecosystems, various types of wastes accumulate in the environment which upset the growth of micro-organisms. A variety of synthetic matter like plastic, rubber, glass, etc., are not degradable, thus, causing environmental problems. Wastes include domestic wastes, industrial and mining wastes and agricultural wastes.

Various types of wastes, if properly managed, can be used and steps have already been taken in this direction. Through gene engineering, new strains of bacteria and other micro-organisms are being bred, which can help to decompose various harmful industrial effluents into non-toxic agents.

Scientists have also developed methods and technology to utilise sewage waste for wood, cattle feed, fertiliser and fuel production. Municipal sewage treatment plants remove large objects, floating matter and settleble solids and also through physico-chemical and biological processes break down the organic matter and remove harmful microbes before release of the treated sewage into water bodies.

Due to the lack of proper facilities sewage is allowed to enter nearby water bodies without any treatment. The untreated sewage causes serious water pollution and also damages the local ecology. It is, therefore, high time that sewage is recycled in an economically feasible and environmentally safe manner to meet the various needs of mankind. Several techniques have been developed for sewage treatment and to extract water for irrigation, fodder, fuel and fertiliser from it. Efforts are also being made to use sewage on a large scale to improve farm lands, reclaim spoilt areas and create new recreational facilities. Use of sewage water for fish farming is common in countries like Germany, Poland, Russia, Hungary, Israel, China, USA, Indonesia, etc. Sewage is being also exploited for fuel, the methane gas released due to bacterial growth can be used as fuel. There is thus an urgent need to develop an efficient waste management system.

Irrigation and Problem of Salinity and Waterlogging

During the last fifty years, vast areas have been provided irrigation facilities. This has definitely increased the agricultural production throughout the world. But, due to the lack of water management and also due to illiteracy of the farmers and the geological structure of a particular region, vast areas are now suffering from the problem of salinity and waterlogging.

Due to the gradual depletion of organic matter and microbial population involved in decomposition, there is increase in salinity and waterlogging, which have adverse effects on soil fertility. Waterlogging and increased salinity are mostly the outcome of poor drainage and mismanagement of water in areas covered by large irrigation projects. The recent example is the Ganganagar and Bikaner districts of Rajasthan, where problem of waterlogging has become serious and several villages have been evacuated and there is no possibility of crop production in these areas. This has happened due to the irrigation water of the Indira Gandhi Canal. In Punjab, the seepage from unlined canals as well as due to irrigation has raised the water table by 7–10 metres above the previous levels. Similar is the problem of salinity, which has damaged the irrigated land. These problems can be checked by proper water management system.

Deforestation and Desertification

The problem of deforestation has arised mainly due to destruction of forests for various developmental or commercial purposes. The growth of population, rapid expansion of cities and other settlements, demand for land for new agricultural farms, mining, construction of roads, dams, etc., have all resulted in mass cutting of natural vegetation. Many useful plants have been made extinct and several thousands are nearing extinction. In the absence of reforestation, the forest cover throughout the world is declining at a rapid pace. The dense tropical forests too are suffering from the problem of mass cutting for commercial purposes.

In order to protect the environment and ecosystem, action plans should be implemented not only to protect the present forest cover but also for large-scale afforestation, so that 30 to 33 per cent of the land area can be developed as natural and/or planted forests. There is also need for the enforcement of present laws strictly to avoid illegal and indiscriminate felling of trees by businessmen, industrialists and rural masses. Both social and agro-forestry should be encouraged and waste lands be developed as forest lands. Besides, public participation should also be encouraged at all levels.

The threat of desertification and expansion of desert area is also due to overgrazing, loss of natural vegetation and other human activities. UNESCO studies have shown that approximately 43 per cent of total land area in the world has been affected by desertification. The suggested measures for the check of desertification are improvement in landuse patterns, stabilisation of sand dunes, development of natural vegetation suitable to arid climate and reduction of population pressure, etc.

Sustainable development should essentially be based on maintaining the fragile balance between productivity functions and ecosystem. For sustainable development, our approach should be holistic on every resource sector and at all levels of local, regional and global scale. Unless we adopt an approach for sustainable development, the problem of environmental degradation cannot be solved.

References

Kayastha, S.L. (1997) Geography and Sustainable Development: Environmental Scenario and Perspectives, *Annals of the NAGI*, 1997, Vol. XVII, No. 1–2.

Pearce, D. (1985) Sustainable Futures: Economics and the Environment, Paper Department of Economics, University College, London.

Ramakrishnan, P.S. (2001) *Ecology and Sustainable Development*, N.B.T., New Delhi.

Redclift, M. (1993) Sustainable Development, *Environmental Values,* 1993, Vol. 2, pp. 3–20.

U.N. Publication, *AGENDA–21, Rio Declaration on Environment and Development,* 1992.

U.N. Publication, *Earth Summit + 5*, 1997.

12

Conservation and Management of Natural Resources

The modern economy is a resource-based economy and the development of a country or a region is based on resources, especially natural resources. Nature has given us abundant resources in the form of water, air, heat, natural vegetation, soil, wild animals, metals, fossil fuels, etc., and man by his technical skill and knowledge is using them in some way or the other right from the dawn of civilisation. 'Resource' is a wide term and its dictionary definitions are:

- upon which one relies for aid, support, or supply;
- means to attain given ends; and
- the capacity to take advantage of opportunities or to extricate oneself from difficulties.

According to Zimmermann (1951), "the word 'resource' does not refer to a thing or a substance but to a function which a thing or a substance may perform or to an operation in which it may take part." In fact, resource is an abstraction reflecting human appraisal and relating to a function or operation. In the process of the utilisation of natural resources man often misuses their accessibility, thus affecting the natural ecosystem. Earlier, this aspect was not given due importance by man, which resulted in the overexploitation of natural resources, but soon he realised that these resources are not unlimited and unless properly used, the day is not far off when he will not only be deprived of them, but will also have to face ecological consequences. This apprehension has given birth to the concept of 'conservation'. In fact, conservation is another form of resource management and resource management is necessary for environmental protection and management. Prior to a detailed discussion on conservation, it is necessary to know the types of natural resources.

Classification of Natural Resources

Any element in our natural environment such as soil, water, forest, wildlife, minerals, etc., that man can utilise to promote his welfare, may be identified as a natural resource. Natural resources vary greatly in quantity, mutability and reusability, therefore, these can be classified into following categories:

Inexhaustible

(a) *Immutable*: Seemingly incapable of much adverse change through man's activities, such as:
   (i) wind power
   (ii) precipitation
   (iii) tidal power
   (iv) atomic energy

(b) *Misusable*: Little danger of complete exhaustion, but when improperly used their resource quality may be impaired, such as:
   (i) solar power
   (ii) atmosphere
   (iii) ocean
   (iv) water power of flowing streams

Exhaustible

These can be maintained by the methods man uses them:

(a) *Renewable*: The living or dynamic resources whose perpetual harvest is dependent upon proper planning and management by man. These are:
   (i) water in place
   (ii) soil fertility
   (iii) products of the land—agricultural products, forests, forage land, wild animals

(iv) products of lakes, streams and impoundments—fresh-water fish, black bass, lake trout, catfish
(v) products of oceans—marine fish, marine mammals
(vi) human resources

(b) *Non-renewable*: Once exhausted, there is no hope of replacement:
(i) mineral resources
(ii) fossil fuels
(iii) species of wildlife

In general, the concept of conservation evolved mainly for non-renewable resources but for renewable resources proper management is also necessary.

Concept of Resource Conservation

The word 'conservation' is derived from two Latin words, '*con*' meaning 'together' and '*servare*', meaning to 'keep' or 'guard'. Literally, therefore, conservation means "to keep together". The word was coined by Gifford Pinchot shortly after the White House Conference of 1908. Some of the definitions of resource conservation are:

> Resource conservation is the scheduling of resource use so as to provide the greatest yield for the greatest number over the longest time period. (P. Haggett)

> Conservation means a sacrifice of the present generation to future generations. Whenever it is carried for, this conflict beginning far before the ideal is reached which conservationists are inclined to advocate. (Ely)

> The optimum allocation of natural, human and cultural resources in the scheme of national development, whereby maximum economic and social security will be assured. (Harold M. Rose)

President John F. Kennedy, in a special message to the US Congress in 1962, interpreted conservation as "the wise use of our natural environment, the prevention of waste and despoilment while preserving, improving and renewing the quality and usefulness of all our resources".

The conservation of natural resources is a concept that deals with the rational use of resources so that a harmony between man's requirements and their availability can be established. While renewable resources need a proper resource management system (a part of conservation), the non-renewable resources require a long-term strategy for their proper use. This is not only necessary to assure long-term resource use but also to protect the ecosystem. Today, every country is very particular about its resources and tries to genuinely adopt conservation policies. During the colonial period between 18th century to mid-20th century, most of the countries of Asia, Africa and Latin America have suffered a lot due to their resource exploitation by colonial powers. Gradually, all these countries attained freedom but either for economic needs or for political adjustment their natural resources have been exploited unwisely, resulting into not only in the loss of several resources but also responsible for many problems of environmental degradation.

The conservation movement, in its present form, was first started in the USA under the guidance of Gifford Pinchot, Head of the US Forest Service from 1898 to 1910, with active support from President Franklin D. Roosevelt. The first wave of conservation was actually started when Roosevelt called the White House Conference on Natural Resources in 1908, followed by a number of important events in subsequent decades, and the enactment of several laws which regulated the manner in which the natural resources of the country can be protected. As a result of the conference, a fifty-member National Conservation Commission comprising scientists, statesmen and businessmen was formed. After 1930, in the USA, several agencies have been established and several acts have been passed to protect forests, soils, wildlife and other natural resources. This tradition was followed by other US Presidents like John F. Kennedy, Lyndon B. Johnson, Richard M. Nixon, Bill Clinton, too, and is still in force to this day. In 1970, Nixon set up the National Environmental Protection Agency.

The conservation wave was not limited to the US only but also started in European countries as well as in Russia and soon all the countries of the world realised the need to conserve natural resources. Therefore, apart from national efforts by each country, efforts on global level have also been started under United Nations. Different agencies of the UN are now playing a key role in setting the world on the long road towards conservation of natural resources and protection of the environment. The increase in public support for conservation measures both at international and local levels has given this movement a new life, but still there is a long way to go in this direction.

Forests Conservation

Where geographical conditions permit vegetation to take the form of trees, the forest is one of the major forms of the natural landscape. The forest resources are valuable as an integral part of the ecosystem, from the commercial point of view, and as providers of shelter to wildlife. Today, forests provide the raw materials for over 5,000 products worth about 23 million dollars. They support industry which employs 1.3 million people. In fact, forests are still the natural habitat of several species of plants and animals, as well as of several tribal groups of the world. But, the most unfortunate setback came in the form of commercial exploitation, which resulted in mass destruction of forest cover year after year. Originally, over two-fifths of the land area of the earth, exclusive of the polar regions, or about 1,200 million hectares, was covered with natural forests. But now more than one-third of this area has been robbed by man and has been turned into barren land.

The history of the exploitation of forests is as old as man himself, but during earlier times it was balanced through a natural growth process because at that time forest cutting was done for personal or community use only. But, with the expansion of agriculture, forest lands have been cleared. More destruction has been done after the industrial revolution and urbanisation. During the colonial period commercial exploitation began and this was the main cause of the depletion of forests. The commercial use of forests today has reached such an extent that it has become a threat to the environment in the form of:

(i) increase in temperature,
(ii) lesser precipitation,
(iii) increased rate of soil erosion,

(iv) increase in frequency and volume of floods,
(v) loss of soil productivity,
(vi) extinction of several species,
(vii) non-availability of several essential forest products, and
(viii) imbalance in ecosystem.

The harmful effects of deforestation are so much that all over the world people and authorities have started realising that forest resources must be conserved properly in order to protect the ecosystem.

The forest is a national resource as well as a social asset. It yields a great social profit which lies wholly outside the realm of business. But, at present, most of the forests of the world are so used that experts predict dire calamities in the not-too-distant future and irreparable damage on a catastrophic scale. If properly used and put on a sustained-yield basis, it will be one of man's greatest resources and for this conservation of forest is the only alternative.

Methods

The following steps should be taken for the conservation of forests:

Regulated and Planned Cutting of Trees

One of the main reasons of deforestation is commercial felling of trees. According to an estimate, about 1,600 million cubic metres of wood have been used for various purposes in the world. Although trees are considered as perennial resource, when exploited on a very large scale, their revival cannot be possible. Therefore, cutting should be regulated by adopting methods like:

(i) clear cutting,
(ii) selective cutting, and
(iii) shelter wood cutting.

The clear cutting method is useful for those areas where the same type of trees are available over a large area. In that case, trees of same age group can be cut down in a selected area and then marked for replantation. In selective cutting only mature trees are selected for cutting. This process is to be followed in rotation. Shelter wood cutting is where first of all useless trees are cut down followed by medium and best quality timber trees. The time gap between these cuttings is helpful in re-growth of trees. In regulated cutting only one-tenth of the forest area is selected for use and rotational system is always followed for their protection. The forest can be managed in such away that a timber crop may be harvested indefinitely year after year without being depleted. This technique is called the 'sustained yield' method adopted by many countries.

Control over Forest Fire

Destruction or loss of forest by fire is fairly common, because trees are highly exposed to fire and once started it becomes difficult to control. Sometimes, the fire starts by natural process, i.e., by lightning or by friction between trees during speedy winds, while in most cases it is also by man either intentionally or unintentionally. According to an estimate, during the period from 1940 to 1950, in the US alone, fires consumed an average of 21.5 million acres of timber yearly and as many as 1,175,664 cases of forest fires occurred during 1955 to 1964 period. In US forests fire is very common and a main cause of the loss of forest. Throughout the world forest fire is common and in most cases they were begun by man. As John D. Guthrie, former fire inspector of US Forest Service, has written: "To stage a forest fire you need only few things—a forest, the right atmospheric conditions, and a spark either from a lightning bolt or a match in the hands of a fool or a knave. The formula is simple—the larger the forest, the drier the air, the bigger the fool, the bigger the fire you will have."

In order to save forests from fire, it is necessary to adopt latest techniques of fire fighting. Some of the fire suppression techniques are to develop three metre wide fire lanes around the periphery of the fire, back fires, arrangement of water spray, fire retardant chemicals should be sprayed from back tank and if possible by helicopters. There must be trained staff of fire fighters to control the fire.

Reforestation and Afforestation

The sustained yield concept dictates that whenever timber is removed, either by block cutting or by selective cutting, the denuded area must be reforested. This may be done by natural or artificial methods. Similarly, any forested land, which has been destroyed by fire or mining activities, should be reforested. In rugged terrain aerial seeding is the method of choice.

Besides all this, fresh afforestation programmes should be started. New plantations will not only increase the forest cover but also help in making up the eco-balance. For afforestation, selection of trees should be done according to local geographical conditions and care must be taken during initial growth of the trees.

Check over Forest Clearance for Agricultural and Habitation Purposes

Most of the present-day agricultural land was once forested and then cleared for the use of agriculture. But now it has reached the stage where further clearance will be dangerous for the entire ecosystem. There are tribals in some parts of Asia, Africa and South America, where shifting cultivation is still a part of their system of land procurement. According to an estimate, about 40 million sq km of land is used for this purpose by 200 million tribals of the world. For the conservation of forests, this should be checked and an alternative method should be devised. Similarly, for the development of villages, towns and cities, forest lands have been cleared and this process continues to this day causing loss of forest cover. This also should be checked and green belts around cities be developed.

Protection of Forests

The existing forests should be protected. Apart from commercial cutting, unorganised grazing is also one of the reasons. There are several forest diseases resulting from parasitic fungi, rusts, mistletoes, viruses and nematodes which cause the destruction of trees. The forests should be protected either by use of chemical spray, antibiotics or by development of disease resistant strains of trees.

Proper Utilisation of Forest and Forests Products

Generally, trees are cut for logs and the rest, including stump, limbs, branches and foliage, etc., is left out as worthless debris.

Further waste occurs at the saw mills. There is thus need to utilise this waste material. Today, several uses have been developed and products like waterproof glues, boards, etc., can be obtained.

Similarly, forests can be used or developed as tourist centres. By using them as tourist centres the country can earn substantial foreign exchange. This practice has been adopted by many countries, both developed and developing. The concepts of 'national park' and 'game sanctuary' have now become popular and every country has developed its unique forest area as a 'national park'. In India alone, there are as many as 92 national parks and game sanctuaries. This is a good method of forest conservation.

Role of Government in Forest Conservation

Although the government of every country is very particular about conservation of its forest resources and has several rules and laws for the protection of forests, but they are not implement in an effective manner. Both national and provincial governments can take some steps in this direction, such as:

(i) pass acts for the conservation of forests,
(ii) survey of the forest resources,
(iii) categorisation of forest areas and proper delimitation of reserved forest areas,
(iv) find out the areas where reforestation can be done,
(v) regulate the commercial use of forest products,
(vi) protect forest from fire, mining and other natural calamities,
(vii) develop national parks,
(viii) encourage forests developmental activities like social forestry, agro-forestry, etc., and
(ix) prepare master plans, both for long-term and short-term period, etc.

Forest Management

Management of forest resources is the key to all conservation efforts. In forest management, the following aspects should be taken into consideration:

(i) survey of forest,
(ii) categorisation of forest,

(iii) economic use of forest,
(iv) administrative setting for forest management,
(v) training programmes for persons engaged in forest conservation activities,
(vi) use of forest land as tourist centres,
(vii) social and agro-forestry,
(viii) development of new techniques for the conservation of forests,
(ix) research for efficient use and conservation of forest, and
(x) policy decisions and their proper implementation.

In brief, conservation of forest resources can be done by cooperative efforts of the government, non-government organisations and the public through a proper management system.

Soil Conservation

Soil is the uppermost layer of earth's surface which supports all forms of life. It is in the soil that all plants have their roots and from which they absorb life-sustaining moisture and nutrients. Man, in turn, feeds directly upon these plants or upon plant-eating animals. Thus, virtually, all terrestrial life ultimately derives from the soil. As a nation's soil resources are fertile and abundant, in like measure will that nation have vigour and stability. When this resource is exhausted because of mounting demands of a swelling population or long mismanagement, the nation's survival is in jeopardy. Some authorities believe that one of the causes of the decline of the Roman Empire was deterioration of soils. The loss of soil, either by natural process or by mismanagement, is harmful not only for agricultural development but to the whole ecosystem; therefore, its conservation has now become a matter of grave concern.

The main reason for the loss of soil is soil erosion, which is of four types:

(i) normal soil erosion,
(ii) soil erosion by water,
(iii) wind soil erosion, and
(iv) accelerated soil erosion.

Normal soil erosion is a natural and a very slow process, and is therefore, not harmful; rather, it helps in soil formation. Soil erosion by water is quite widespread and a main reason of soil loss and loss of fertility. Water soil erosion is of four types, viz., sheet erosion, rill erosion, gully erosion and riparian erosion. The third agent of soil erosion is the wind. Winds are more effective in arid and semi-arid regions. While the accelerated soil erosion is done by human activities, which accelerate the process of soil erosion by deforestation, grazing, mining, defective agricultural practices, etc., landslide and leaching also cause soil loss. The problem for soil loss now has become a worldwide problem and its conservation is necessary.

Methods

Soil erosion can be controlled by adopting land management practices and also by changing the pattern of some human activities which accelerate soil erosion. The following techniques can help in controlling soil erosion:

(i) *Contour farming*: Contour farming may be defined as ploughing, seeding, cultivating and harvesting across the slope, rather than with it. Thomas Jefferson, who wrote in 1813, "we now plough horizontally, following the curvature of the hills ... . Every furrow thus acts as a reservoir to ... retain waters ... scarcely an ounce of soil is now carried off."

(ii) *Strip cropping*: On land with a dissected slope, planting crops on contour strips will be an effective erosion deterrent. For effective control the width of the contour strip should vary inversely with the length of the slope. Strip cropping should be combined with crop rotation, so that a strip planted to a soil depleting, erosion-facilitating corn crop one year will be sown to a soil-enriching and protecting strip of legumes the next year.

(iii) *Terracing*: The practice of terracing has been common in ancient China. The flat, step-like bench terraces are now not useful. The modern terrace is an embankment of earth constructed across a slope in such a way as to control water run-off and minimise erosion. To be effective, terrace must check water flow before it attains sufficient velocity to loosen and transport soil.

(iv) *Gully reclamation*: Gullies are danger signals that indicate that land is eroding rapidly and may become a wasteland as in the case of vast areas along the rivers Chambal and Yamuna. If a gully is small, it may be ploughed in and then seeded to quick growing crop like barley, maize, jowar and wheat in order to check erosion. In case of severe gullying, small check dams of manure and straw constructed at 5 metre distance may be effective, because silt will collect behind the dams and gradually fill in the channel. Earthen, stone and even concrete dams may be built along the gully. Once dams have been constructed and water run-off has been restrained, soil may be stabilised.

(v) *Shelter belts*: These are the 'green belts' of trees which help to break the force of strong winds and thus prevent or cut to a minimum the blowing away of the loose topsoil. In areas where wind erosion is more, rows of trees may be helpful to check the flow of winds. Apart from this, these trees will also add colour to the landscape and help to control the desert spread. Soil blowing away can also be controlled if local shrubs and small trees planted in a systematic way. Even useful trees can be planted and harvested after a regular interval of two to three years.

Other methods of soil conservation are:

(i) expansion of vegetative cover and protective afforestation,
(ii) controlled grazing,
(iii) flood control,
(iv) prohibition of shifting cultivation,
(v) proper land utilisation,
(vi) maintenance of soil fertility,
(vii) land reforms, reclamation of wasteland,
(viii) establishment of soil research institute and training of soil scientists, and
(ix) effective agencies for soil management.

Conservation of Water

Water is an indispensable component of our ecosystem and the essence of our life. On an average, each individual uses 50 litres of water per day in rural areas and 150 litres in urban areas. Apart from drinking purposes, water is used for irrigation as well as in industries. Wherever available, about 12,000 m^3 to 14,000 m^3 water is used annually to irrigate one hectare of land. The main source of water is rainfall occurring through hydrological cycle, while other sources include underground water and water from the melting of snow. The oceans are full of water but because of salinity this water is not usable, so also with several inland seas. In fact, usable water is very limited and there are areas in the world where water is a scarce commodity. The unfortunate situation is that wherever water is available, users never think about its proper use and they not only misuse it but waste it too. Therefore, conservation of water is an absolute necessity for all countries. A few steps in this direction are:

Proper Distribution System of Water

The water distribution system should be such that loss of water should be minimum and it should remain free from pollution. According to local/regional conditions, the distribution system should be developed. A few general steps for distribution of water are:

(i) As far as possible water distribution should be done through pipes so that loss will be minimum and water will remain free from pollutants.
(ii) If water storage place is small, it should be cemented and if its size is big, the selection of site should be done where non-porous rocks exist.
(iii) The channels and distributors of canals should be cemented, and in fields also either pipes be used or cemented channels (*nala*) be constructed.
(iv) For irrigation, sprinkle and drip system be used.
(v) If possible, water tanks should be covered to avoid evaporation and also for cleanliness.

Rational Use of Underground Water

Underground water is also a main source of water, specially in arid and semi-arid regions. The overexploitation of underground water often results in the lowering of water table, intrusion of saline water, subsidence of land, etc. The prime need is the proper and limited use of water and also geological and hydrological survey of the region for assessment of water availability.

Purification of Waste Water

The system of the recycling of water should be used so that waste water could be used for irrigation purposes. There are several mechanical devices for waste water treatment. The government agencies and industrial establishments should establish treatment plants. This will also help in control of water pollution.

Control over Deforestation

The loss of vegetation is also a cause of drought and reduction in rainfall and indirectly a cause of lowering of the water table. In order to maintain humidity in the atmosphere which helps in rainfall and to minimise evaporation rate, forest cover should be maintained.

Protection of Water from Pollution

Apart from the availability of water, quality of water is also important; in other words, water should be free from pollution. The nature, form and types of water pollution and its effects have been discussed in detail in Chapter 8. Today, several techniques such as physico-chemical purification methods—hydrolysis, electrolysis, ion-exchange, absorption, congutation, chlorination, ozonisation, etc., are in use. Similarly, biological purification method is also used.

The conservation of water can be done only by proper water management system. Apart from the above mentioned measures, water can be obtained from desalination of sea water and even by artificial raining. Besides, mutual cooperation in the use of water resources is also necessary.

Wildlife Conservation

Like forests, wildlife is also a national resource, which not only helps in maintaining the ecological balance but is also beneficial from economic, recreational and aesthetic points of view. There was a time when human interference was minimum, the number of wild animals were quite high and there was no problem of their protection or conservation. But, with the expansion of agriculture, settlement, industrial and other developmental activities and mainly due to greed of man, the number of wild animals gradually became lesser and lesser. With the result that several species of animals have become extinct and several others are on the verge of being so. Deforestation is also one of the main reasons for the loss of wildlife. Mass killings of wild animals for their meat, bones, fur, teeth, hair, skin, etc., are going on throughout the world. Therefore, the need for wildlife conservation has now become a necessity.

Population growth, expansion of agriculture and livestock raising, building of cities and roads, and pollution are among the many pressures on the natural habitat of wildlife. Along with illegal hunting, habitat reduction and its degradation has threatened the bio-diversity of the regions where these are rampant. Preservation of wildlife does not mean a blanket protection to all faunal and floral species; rather, it implies a proper, judicious control over the multiplication of plants and animals, which interact together to provide a proper environment to man whose very existence is in peril today. Due to the irrational use of natural and biotic resources of the earth in the past, most of the wildlife has been destroyed beyond retrieval. It is our urgent duty to protect the natural splendour of ecosystems and to evolve a system of co-existence with every living creature upon the earth.

Although must countries of the world are very particular regarding conservation of wildlife, the number of wild animals is reducing day by day. World Wild Life Fund is the international agency, which is doing commendable work in promoting the protection of wildlife. There are national agencies also engaged in the conservation of wildlife. Some steps in the direction of wildlife conservation could be as follows:

(i) To survey and collect all the information about wildlife, especially, their number and growth.
(ii) To protect habitat by protecting forests.
(iii) To delimit the areas of their natural habitat.
(iv) To protect wildlife from pollution and from natural hazards.
(v) To impose complete restriction on hunting and capturing of wildlife.

(vi) To impose restrictions on export and import of wildlife products and severe punishment to be given to those who indulge in this activity.
(vii) To develop game sanctuaries for specific wild animals or for general wildlife.
(viii) To make special arrangements to protect those species whose number is very limited.
(ix) To develop general awareness at national and international level regarding protection of wildlife.
(x) To adopt a system of wildlife management through trained personnel.

India is a good example where several steps have been taken for wildlife conservation. It is a country of varied wildlife, where more than 500 types of wild animals, 2,100 types of birds and about 20,000 types of reptiles and fishes have been found. According to an estimate, in India, about 200 species of wild·animals and birds have already become extinct and another 2,500 are on the verge of extinction. Some of them are black buck, chinkara, wolf, swamp deer, nilgai, Indian gazelle, antelope, tiger, rhinoceros, gir lion, crocodile, flamingo, pelican, bustard, white crane, grey heron, mountain quail, etc. In India, the government and NGOs are taking keen interest in the protection of wildlife. The Wild Life Protection Act, 1972 has several provisions for the conservation of wildlife. As many as 165 game sanctuaries and 21 national parks have been developed to protect the natural habitat and wild animals. Apart from this, a Wild Life Conservation Week is also celebrated from 7th of October every year. But still there is a long way to go in this direction.

Conservation of Energy Resources

Almost 90 per cent of the world's current energy supply is based on fossil or mineral sources, oil, coal, gas and uranium. Only 10 per cent is from renewable sources, especially hydropower and biomass, while the share of solar and wind energy is below 1 per cent. Table 12.1 indicates the pattern of primary consumption in the world.

Table 12.1
Primary Energy Consumption in the World

Energy Source	*Worldwide Consumption*	*Percentage*
Oil	21678.10^6 barrels	34.2
Coal	4765.10^6 tonnes	30.2
Gas	1923.10^9 cbm	19.1
Nuclear	1670 TWH	5.0
Hydropower	2050 TWH	6.1
Wood/biomass	1219.10^6 tonnes	5.3
Wind	3 TWH	1.10^{-2}
Solar	0.1 TWH	1.10^{-3}
Geo-thermal energy	76 TWH	0.1

It is undisputed that this structure of energy supply involves three problems: if consumption rates do not change, reserves of non-renewable energy will largely run out in one or two generations; the waste products caused by consumption in the water, the earth and in the atmosphere threaten mankind and the environment; and finally, the cost of this shortage and the effects on consumption will be too much. The use of energy is increasing day by day and according to an estimate the oil and gas will be in shortage after 2025 AD and coal reserves will also fall short after 2200 AD. Even today, there are problems in coal mining. The world supply of natural gas is also limited.

Nuclear power, which amounted to over 5 per cent of global energy production, accounts for very limited use and after the Chernobyl accident more technological precautions are needed. At present, there is need for adopting methods for conservation of energy resources and also to develop and use more and more renewable energy sources.

The conservation methods should be applied for coal, petroleum and natural gas in order to preserve these resources, so that their use could be possible for a longer period and their environmental effects could be minimised. The coal mines have become deeper in most of the countries, therefore, new techniques should be applied. The new methods like gasification may help in deeper mining operations. The coal should be classified and cleared at production centres itself. The use of secondary coal products like coaltar, ammonium sulphate, gas, light oil, benzol, sulfur, etc., should be encouraged. A better transportation and marketing system as well as research in the use and mining of coal can also be beneficial.

The conservation of petroleum or mineral oil is the need of the times because the availability of petroleum reserves is limited and the use of oil is increasing day by day, especially in the field of transportation. The following steps may help in conservation of petroleum:

(i) reduction of the consumption of petroleum,
(ii) development of such engines which consume lesser petroleum,
(iii) intensive survey for new resources,
(iv) exploitation of oil resources from ocean floors,
(v) correct estimation of available resources,
(vi) new techniques for oil drilling and refining should be developed,
(vii) to reduce petroleum consumption in certain areas such as for military purposes,
(viii) to develop alternative source of energy for automobiles, etc.,
(ix) international cooperation in reduction of the consumption of petroleum products, and
(xi) research institutes working in this field should intensify their work.

The energy crisis can be solved to some extent by proper use of renewable sources of energy such as solar energy, wind energy, tidal energy, hydropower, ocean energy and geo-thermal energy. Renewable energies are also called the energies of the future, because in principal they are inexhaustible and neutral in terms of CO_2. The renewable energy sources utilised by man come from three primary sources:

— isotopic disintegration in the earth's interior,
— thermo-nuclear conversion (fusion) in the sun, and
— the movement of planets, combined with mass attraction.

The greatest of the renewable energy sources is *solar energy* or solar radiation reaching the earth's atmosphere year after year. Every 29 seconds the solar energy that falls on our planet is equivalent to human energy needs for one day at the 1980 consumption level. In case of solar-thermal utilisation, the solar energy is converted to heat through absorption. This can occur through passive systems, i.e., through constructional measures for collecting, storing and distributing solar energy which renounce thermal installations to a considerable degree or through active system, i.e., collectors or absorbers. The advantages of using solar energy include limitless supply, no air pollution, no water pollution, no harmful wastes, no possibility of a large scale explosion or disaster. Thus there is a need to evolve cheaper technology to harvest solar energy at least for tropical regions as it is the greatest source of energy.

There is no need to stress the importance of *water energy* or hydropower. Water energy is already in use in developed countries, but in developing countries vast energy potentials exist.

There is enough *wind energy* available to supply the world's energy needs, and it is technically feasible to build windmills capable of producing electricity. These can be built either as small home-sized units or as large central generators. Besides the conversion of rotation energy into electrical energy, the conversion of rotation energy into mechanical energy also finds application in pumping of water, etc.

Ocean energy refers to a number of different renewable energy sources—tidal energy, wave energy, ocean currents and ocean heat. Today, the significant attempts at harnessing the energy of the ocean have been related to tidal power. In many coastal regions, the flow of the tide naturally funnels through narrow entrances into bays and estuaries, and strong currents can be used to generate power.

Energy derived from the heat of the earth's crust is called *geo-thermal energy*. In various places on the globe, hot springs, geysers and underground steam are available, which can be used for power generation.

Bio-energy can be produced by organic compounds through chemical and biological processes. The combustion, gasification and liquification helps in production of gas which can be used as a source of energy. The bio-gas plant has now become a common feature in rural India.

Viewed in the long-term perspective utilisation of renewable energy sources is of major significance, in terms of both the conservation of the reserves of fossil fuels and the reduction of pollutant emissions caused by the conversion of energy from coal, oil and gas. The renewable energy sources have good chances for the future energy supply. How large their contribution will be, depends upon the question of how safe, economical and environmentally friendly each particular 'new' energy is.

Reference

Zimmermann, E.W., *World Resources and Industries*, Harper and Row, London, 1951, p. 7.

13

Population, Eco-crisis and Sustainable Development

Ecologists have defined population as "a collective group of organisms of the same species or other groups within which individuals may exchange genetic information occupying a particular space" (Odum, 1971). Thus, each and every organism is an integral part of the ecosystem and has its impact over it. In the present context, population means *human population*, which not only have access to the environment as habitat but through its multifarious activities also changes certain natural processes in the ecosystem, thus, becoming a cause of the ecological crisis. The situation becomes more alarming due to the population pressure and overexploitation of environment.

The interaction between population and environment is very complex and dynamic. This complicated interrelationship is not yet fully understood or spelt out at microlevel and spatio-temporal variations over the nation or the globe understood in its entirety. But, there is no doubt about the fact that the environment is constantly changing due to human activities, which leads to problems like soil erosion, devastating floods, acute droughts, unforeseen climatic changes, desertification and general degradation of the environment. Figure 13.1 indicates the relationship between population and eco-crisis.

The environment—natural, economic and cultural—acts and reacts in a composite manner and determines the growth and spatial pattern of the population. On the other hand, human activities create chain reaction in the ecosystem. These reactions, pollution

Figure 13.1
Population and Eco-crisis

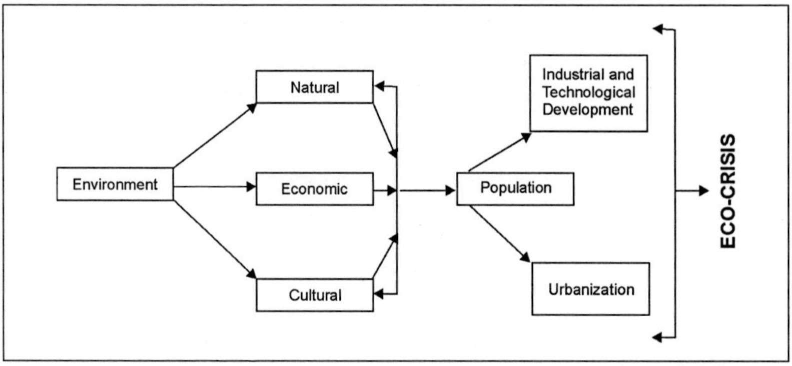

and disturbances are now evoking great concern. All these aspects of the population and its impact over the environment need analysis so that this problem can be understood and steps suggested for future planning.

Population Growth and Distribution

The growth of human population is a continuous process and in the last 10,000 years the population of the world has increased over a thousand-fold, to about 4.5 billion, and much of that change actually has occurred in the last few hundred years. The current population of the world is estimated to be over 6.1 billion. While Table 13.1 indicates the growth of population, Figure 13.2 shows the trend in growth of world population, which is higher in developing countries.

According to United Nations estimates, the population growth rate at present hangs around 2.5 per cent annually. The rate of increase is staggering. Every 30 seconds, about 117 live babies are born and some 46 people die, for a net increase of 71 new inhabitants. This means, over 2,00,000 new mouths to be fed each day or nearly 75 million a year. In spite of the implementation of population control programmes, in the next few decades, there will be high growth rate, because of the effects of declining fertility. After that, if current programmes of family limitations succeed, there are reasons to believe that growth rate will slacken and the population will stabilise in the distant future. The growth rate is much higher in developing countries than in developed countries as shown in Figure 13.2.

Table 13.1
Growth of World's Population (1650–2001)

Year	*Population (in millions)*	*Average Annual Growth Rate (in millions)*	*Percentage*
1650	545	1.8	0.29
1750	728	3.5	0.44
1800	906	5.3	0.51
1900	1,171	8.7	0.64
1920	1,810	21.8	1.11
1940	2,246	24.7	1.05
1950	2,493	49.1	1.81
1960	2,984	64.8	2.15
1969	3,788	145.8	2.23
1981	4,492	70.4	1.7
1990	5,293	80.1	1.76
2001	6,137	84.4	1.6

Figure 13.2
Trends in Growth of World Population

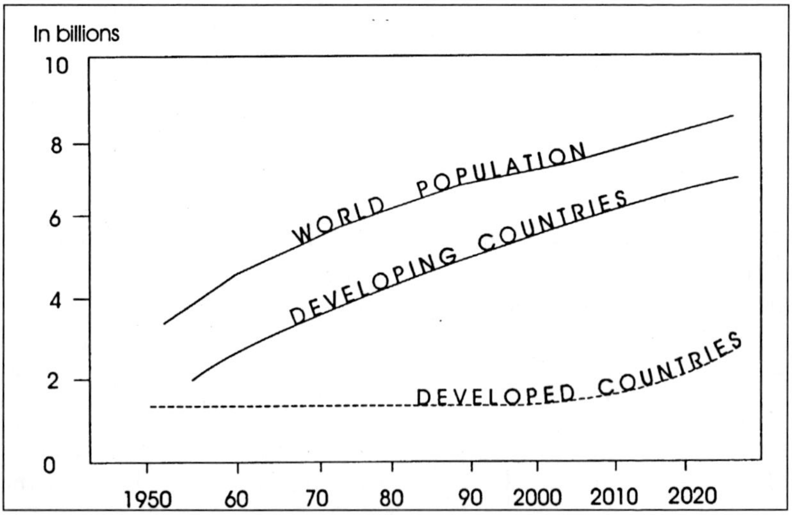

Another aspect of the population, which is considered to be effective in terms of its impact over the ecosystem, is its spatial patterns, both at macro and micro levels. The characteristic feature of the distribution pattern of world population is the concentration of population in limited areas. The continent-wise population is given in the following table:

Table 13.2
Population of the Continents

Continents	*Population (in millions)*
Asia	3913.84
Africa	891.43
Europe	729.34
North America	512.42
South America	371.27
Australia	327.44
Antarctica	Nil

This unequal distribution of population is the main cause of several environmental problems. On the one hand, there are countries like China and India, where population pressure is maximum and resources are limited, and on the other hand, in countries like USA and Canada population is less than resources. In fact, the positive correlation between population and resources is a key component for sustainable development.

The Demographic Transition

Generally, nations go through a four-stage change process over time known as the 'demographic transition' as depicted in Figure 13.3.

Stage I (no growth): Both birth and death rates are high, resulting in little or no population increase. There are no clear examples of such nations in present times.

Stage II (rapid growth): The birth rate is still high, but the death rate is falling due to advancement in public health. Many

Figure 13.3
Demographic Transition Model

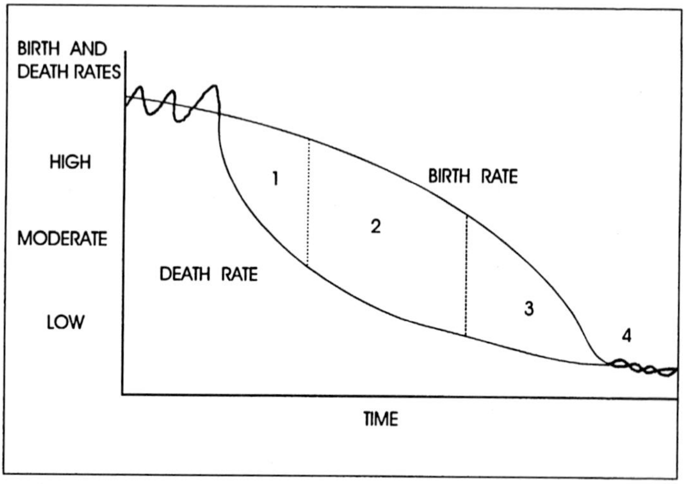

developing countries are in this category, prominent are, India, Bolivia and Saudi Arabia.

Stage III (slow growth): The death rate is relatively low and the birth rate is declining such as in China and Indonesia.

Stage IV (stagnation): Birth and death rates are both low, giving a very low rate of increase. Most of the European countries come in this category.

In fact, the rapid growth of population leads to chronic problems like malnutrition, unemployment, disease and, of course, degradation of environment.

Ecological Checks on Growth of Population and Carrying Capacity of Environments

Although growth in population is a continuous process, there are certain natural checks, which can have limiting effects over unlimited growth. The famous demographer, Thomas Robert Malthus, in his writing *Principles of Population* (1798), had pointed out the dire ecological consequences in the continuing growth of population. He claimed that population has a tendency to increase geometrically (in series 1, 2, 4, 8, 16, ...), while the food resources for that population, even with improving agricultural methods, increase only arithmetically (in series 10, 20, 30, 40, 50, ...). When growth reaches that point, it can be kept in check, according to Malthus, only by "war, vice and misery". But, in 1817, Malthus paid more attention to the curtailment of population increase through birth control than to the gloomy devices of war, vice and human misery.

The carrying capacity of environment means the maximum number that the local environment can afford; or the biological capacity to provide food to them. This is represented in Figure 13.4 by three graphs illustrating hypothetical relations.

When population growth approaches a point of ceiling, the three situations are: first, the rate of increase may be unchanged until the ceiling is reached, and then abruptly drop to zero; secondly, the rate of increase may decline; and thirdly, the population may overshoot the ceiling.

The three possible adjustments are instantaneous adjustment, progressive adjustment and progressive approximation. In fact, the carrying capacity of the environment also changes with time due to technological and scientific developments.

The gist of Malthus' thoughts is that growth of population, if beyond the level of saturation, makes adjustments in its own way. We can say that growth of population will be steady and smooth only when there is an adjustment between environment and growth. Increase in population leads to overexploitation of environment and becomes a cause of eco-crisis, ultimately leading to large-scale deaths due to famines, floods, earthquakes, landslides, etc., and also sometimes due to epidemics. The wars between nations and other conflicts are also sometimes due to overpopulation.

In the history of mankind, there are several instances when the population has been decreased due to natural calamities. The death toll in India due to famines of 1837, 1863 and 1876–78 was 8, 10 and 50 lakhs respectively. In China, during the 1877–79 famine, 90 lakh lives were lost and another 30 lakh in 1928–29. Even in Soviet Union, in 1932–34, 40 lakh people died due to famines. According to Malthus, "famine seems to be the first, the most dreadful

Figure 13.4
Environmental Constraints on Growth of Population
(Three Hypothetical Relations)

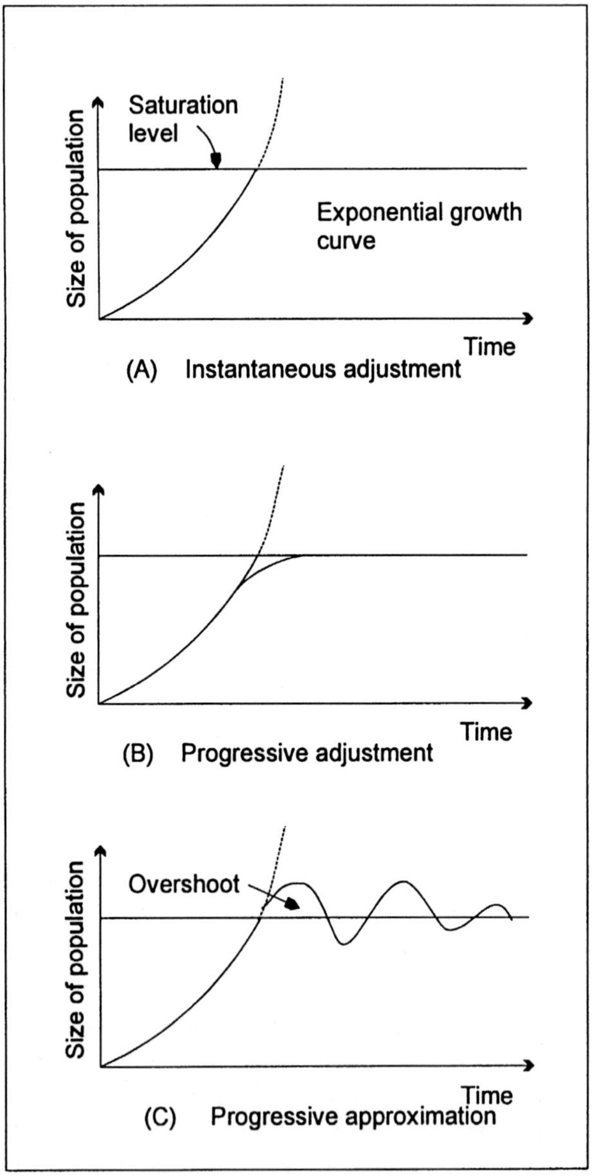

resource of nature. The power of population is so superior to the power in the earth to produce subsistence for man that premature death must in some shape or other visit the human race".

Similarly, floods, earthquakes, volcanic eruptions, landslides, storms, etc., have claimed lakhs of lives every year. Apart from natural hazards, man himself acts as a destroyer either for his selfish motives or for political and economic superiority. The First World War (1914–18) claimed 72 lakh lives and in the Second World War (1939–45) the number was 73 lakhs. In the Spanish civil war (1936–39) 63 lakh people died. While in India communal riots of 1946–48 claimed 59 lakh deaths, American civil war (1861–65) 58 lakhs, Russian revolution (1918–20) 57 lakh, etc. The Vietnam war, Arab-Israel conflict, India-Pakistan wars, India-China war, Gulf war, Iraq war, etc., are so many examples in which thousands of lives have been lost. The growth of population may not be a direct cause of it, but it is certain that population pressure resulted in the social and economic tensions and became a cause of environmental degradation.

Population and Food

Food is a common denominator of life. But for many people on the earth it is not available. Thousands of people are dying of hunger and millions are suffering from malnutrition. There is a close relationship between population, food supply and environment. The production of food, not only of farm products but also marine and others, depends on suitable climatic conditions. The non-availability or limited supply of food is a cause of tension and conflict on the one hand, and overexploitation of resources, on the other.

The world food supply, like population, is unevenly distributed. Most of the developed countries either have less population and surplus food supply or are endowed with enough resources to obtain food from outside. On the other hand, in developing countries, food supply is limited and population is more resulting in hunger and insufficiency of essential nutrients causing infant mortality, impaired development and disease.

The most important question related with population and food supply is the nutritional value of the food measured in terms of calorie intake. The UN World Food Survey has pointed out that 2,550–2,650 calories is the minimum requirement for each individual. A balanced meal means:

(a) 1,200 to 1,800 calories from cereals,
(b) 100 to 200 calories from starch,
(c) 100 to 150 or 200 calories from fats,
(d) 250 to 300 calories from meat and pulses,
(e) 100 calories from fruits, and
(f) 300 to 400 calories from milk and milk products.

If we look at the pattern of the world calorie intake status (as shown in Figure 13.5), it becomes clear that most of the countries of Asia, Africa and Latin America have very low, i.e., less than 2,500 calories intake per day per person. Some of the countries have a medium level intake (2,500 to 2,900 calories) and only USA, Canada, countries of Europe and Russian Federation, Australia and New Zealand have a high calorie intake food. The pressure to produce food caused by a growing population or by crises such as droughts or floods can result in long-term damage to environment.

John Bongaarts, in March 1994 issue of *Scientific American,* stated:

> Feeding a growing world population a diet that improves over time in quality and quantity is technologically feasible. But, the economic and environmental costs incurred through bolstering food production may well prove too great for many poor countries. The course of events will depend crucially on their governments' ability to design and enforce effective policies that address the challenges posed by mounting human numbers, rising poverty and environmental degradation. Whatever the outcome, the task ahead will be made more difficult if population growth rates cannot be reduced.

Urban Ecology

Rapid utilisation is a worldwide phenomenon. According to an estimate, in 1935, the number of "million cities" was 56, which has now crossed 125. In 1800, only 1.7 per cent population was living in

Figure 13.5
World Pattern of Calorie Status

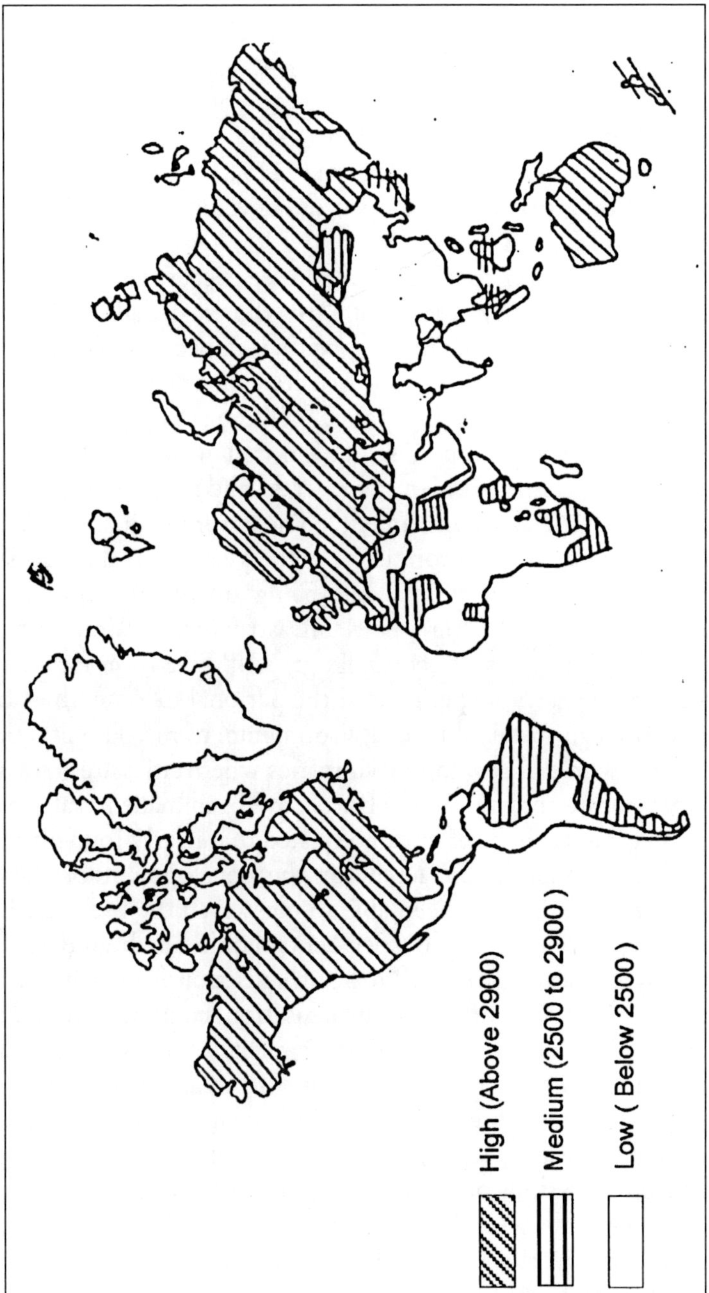

cities, which raised to 5.5 per cent in 1900 and 25.8 per cent in 1975 and by year 2010 it will be more than 40 per cent. Trends in the rate of urbanisation vary from country to country. For example, the United States has more than 70 per cent urban population, while in India its percentage in 2001 was 27.78. The pattern of urban population growth in developed and developing countries has been shown in Figure 13.6.

Urbanisation continues to increase because the benefits gained from high density crowding are the so-called agglomeration economies. While having a positive impact on income levels, employment, production and market economics, urbanisation has brought with it many problems, including a shortage of housing, inadequate water supply, sanitation and waste disposal facilities, congestion, traffic problems, air, water and noise pollution, as well as unsafe social environment. The overall quality of urban environment has deteriorated over the years with the largest cities reaching saturation points and being unable to cope with the increasing pressure on their infrastructures. While cities are considered 'engines of economic growth' given the fact that they contribute nearly 60 per cent of the national income, they are also inherently unsustainable in environmental terms. The situation has become worse in developing countries where cities are growing at a very fast rate and very little efforts have been made to safeguard the urban ecology. In brief, it can be stated that nearly four-fifths of the total investment is cornered for the development of the urban habitat, yet it has not been able to meet even the basic needs of its dwellers. About 30 to 60 per cent people are bound to live in clumsy houses and slums. Now, urbanisation is posing a threat to environment. The process of urbanisation, coupled with industrialisation, has caused urban atrophy, urban congestion, growth of stinking slums, and impurification of air and water. The urban slum people are denied minimum shelter space and even minimum amount of water, sanitation and electricity. There is a need to develop new kinds of technology to reconstruct industries in order to protect urban ecology. The process of urbanisation should be channelised in such a way that the entire urban area will remain pollution-free and become an environment-friendly zone.

Figure 13.6

Actual and Projected Urban Population Growth (1950-2025)

Actual and Projected Percentage of Urban Population in Developing Regions

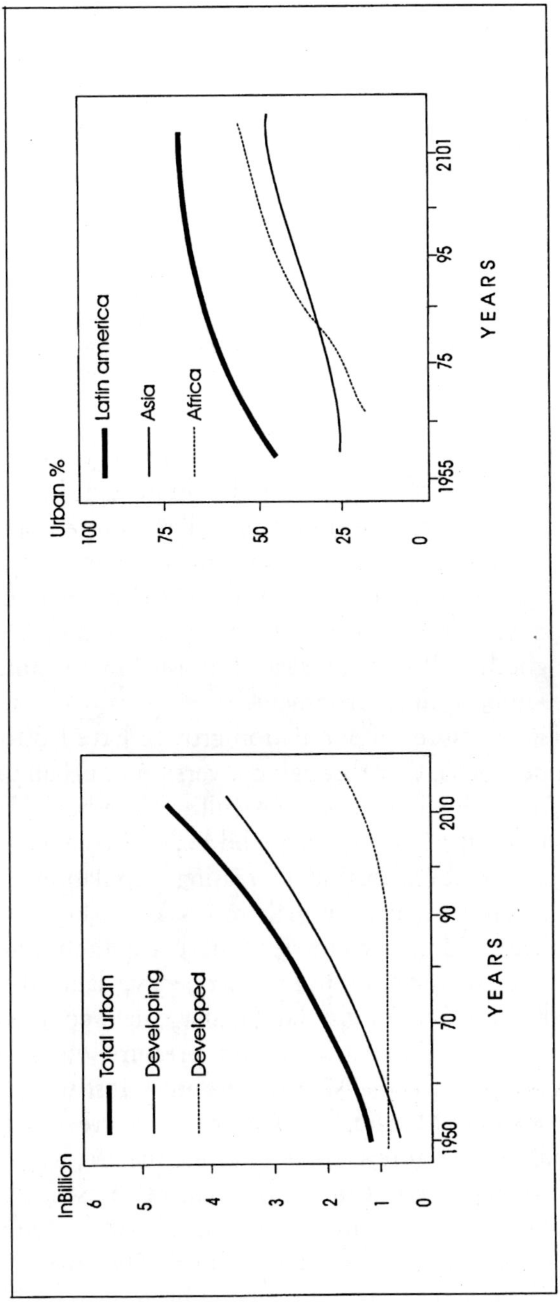

Effects of Population Growth

According to an estimate, over 8 million people are added to world's population every year. That means more food must be produced, more water for drinking and agricultural purposes, more materials to provide shelter, and so on. Bigger population has put bigger pressure on environment and increased the demand for earth's resources. The collective impact of over 6 billion people living on earth is tremendous. Besides, the geographic distribution of people, the concentration of people in urban areas and the demographic characteristics of regional populations also have an important influence in the effects of human activity on the environment.

As described earlier, there is great regional disparity in population size, which, in turn, has important implications for the environmental impact of human activity. More than 90 per cent of population growth is among less developed regions. As the residents of less developed countries embrace the living standards of the industrialised world, per capita consumption of energy and other goods will also increase at a faster rate resulting in a much greater impact on environment.

The pressures of population growth have led to force the loss of prime lands either through conversion to urban uses or through degradation by imprudent agricultural methods. In many cases, human activity has degraded land beyond recovery. A number of studies have revealed that increasing population and population density have negative impact on soil quality. A study of 38 sub-Saharan African countries found that the population growth tended to increase cropping frequency, land degradation and breakdown in community land management controls.

Increasing population exerts pressure on forest lands too. Recent studies by the World Resources Institute and the World Conservation Monitoring Centre have revealed that nearly one-half of the forests that covered the earth have now gone. Tropical forests constituting one-half of the world's forest cover, declined by about 450 million hectares between 1960 and 2000.

Forest loss leads to habitat loss. The earth has around 14 million different species, of which only about 1.7 million have been identified and described. Many of these species are threatened with extinction primarily because of habitat loss. An estimated 1 to 11 per cent of the world's species are facing the threat of extinction. Population growth and density contributed to deforestation in many countries.

Population growth also results in consuming metal and minerals in larger quantities. Global consumption of iron and other metals rose more than three times during the last four decades because of growing demand of goods. Population growth has also led to massive energy consumption. Energy consumption increased to 144 per cent in Africa and 185 per cent in Asia during the period 1975–2000. Oil reserves are depleting and new oil reserves have hardly been discovered in recent years.

Water available for human use is also shrinking because of pollution from agriculture and other human activities. Underground deposits of water are being pumped out faster than it is being replenished by surface water. Safe drinking water is still not within the reach of most of the people in many geographical regions. About 1.2 billion people in developing countries lacked safe drinking water supplies and about three billion people lacked access to sanitation services by the end of last century.

Population density plays a vital role in environmental degradation. Due to migration from rural to urban areas, 90 per cent of future population growth and a major share of economic growth is expected to occur in cities. Urban growth leads to encroachments of farmlands, destroys habitats and threatens sensitive ecosystems and inshore fisheries. It also leads to improper atmosphere that promotes environmental degradation. In brief, population pressure leads to several environmental problems as shown in Figure 13.7.

Population and Sustainable Development

Both population and environment cannot be viewed in isolation and their impact on development is also apparent. Actually, population, environment and development should be treated as interacting aspects of the global phenomena. Several aspects of population growth and its impact on environment have already been discussed. Similarly, development is done at the cost of natural

Figure 13.7
Population Pressure and Environmental Problems

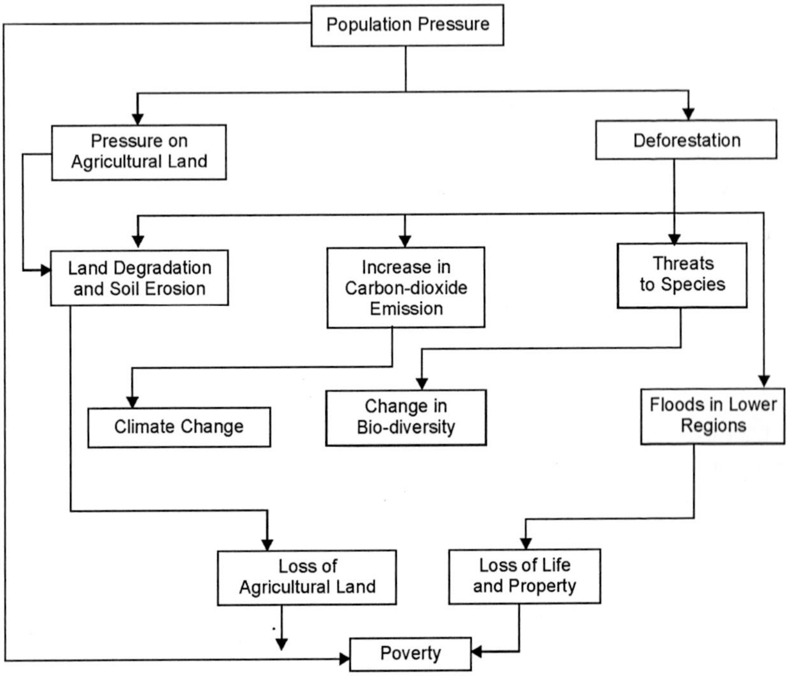

resources and is responsible for the degradation of environment. It is also true that development is a continuous process. But, at the same time, it is also necessary to create a balance between growth of population, development and quality of environment.

The United Nations, realising that the environment was under great stress, and air and water pollutions were on the increase and the human environment was in a bad shape, particularly in developing countries, where two-thirds of the population lived in poverty, organised a World Summit on Human Environment in 1972 in Stockholm. This was the beginning of the worldwide awareness towards the various environmental problems, including problems related to population growth and its impact on environment. The environmental problems are often related with poverty and the problems could be overcome by development. The

United Nations established a World Commission on Environment and Development in 1983 which published its report under the title 'Our Common Future' in 1987. The report considered the problems of sustainable development, food security, the urban challenge, development and the environment. In this direction U.N. Conference on Environment and Development at Rio de Janeiro was organised in 1992. In this conference, *Agenda 21* was adopted, alongwith several issues, emphasis was also placed on sustainable development.

Similarly, the World Population Conference, held in 1974 in Bucharest, recommended that the developing countries should reduce the rate of population growth and modernise methods of family planning. After a decade of Bucharest conference, the second World Population Conference was held in Mexico in 1984 with a different outlook. The conference emphasised that the development should be linked with population, physical resources, environment and the standard of living. Family planning may not be fully effective unless empowerment of women is undertaken and achieved with the sustained development of rational utilisation of resources. In 1994, another World Population Conference was held in Cairo, in which emphasis was given on women's empowerment, which would help in reducing the rate of growth of population and also in protecting environment.

The limits of growth is a subject of debate. A group of intellectuals of the Club of Rome undertook a project called "The Predicament of Mankind" and its report prepared in 1972 under the title, "the Limits to Growth". The report concluded that the continued growth would lead to exhaustion of natural resources, collapse of industrial and agricultural production owing to shortage of resources and fertilisers and a decline in population caused by famines and epidemics. It was also stressed that reduction of production and pollution would not solve the problem of exponential growth of population. The only alternative according to report is a "no growth" policy for both industry and population. The population growth should be restricted to 'zero' level and development be done with minimum damage to environment.

Critical Issues

Some of the critical issues related with the problem of population and eco-crisis are as follows:

(i) Both the population and environment are complex systems and their interactions are also dynamic, complex and inconsistent with time and space, therefore, planning for each country/region should be based on realistic research.
(ii) The production systems should be taken into consideration as interaction between population and environment varies with production system, i.e., capitalistic, socialistic or mixed.
(iii) The high population growth and high fertility among the poor masses leads to more pressure upon the world's resources and deterioration of the environment.
(iv) The migration of population from one natural environment to another or from rural to urban areas is responsible for rapid environmental degradation.
(v) Along with environmental issues, proper attention should be given to relieving hunger, poverty and malnutrition. Only then the quality of environment can be improved in developing countries.
(vi) For growing population a three-fold development policy should be adopted:
    (a) Social welfare (food, water, shelter, health and literacy);
    (b) Self-reliance (technological, economic and strategic); and
    (c) Advanced science and technology (frontier level areas).

    This can be achieved through:

    (a) strategy for resource allocation,
    (b) manpower development,
    (c) extension and transfer facilities, and
    (d) technological management.

Indian Scenario

India is the best example of the relationship between population and environment. Right from ancient times too much importance has been given to nature or environment. As stated earlier, the environment not only directs but also determines the growth of population, its distribution and functional structure. There was complete coordination and co-existence between man and nature but with the rapid growth of population this balance has not only been disturbed but has also become a threat to the ecosystem and is now responsible for the degradation of the environment.

India's population has very rapidly grown since the demographic watershed of 1921, when medical treatment, sanitary improvement, etc., started reducing mortality drastically and led to concomitant population explosion. In 1921, there were about 319 million Indians, which rose to 842 million in 1991 and 1,027 million in 2001, as shown in Table 13.3.

Table 13.3
Growth of Population in India

Census Year	*Population (in millions)*	*Percentage of Decadal Variation*
1901	238.39	—
1911	252.09	+ 5.75
1921	251.32	- 0.31
1931	278.97	+ 11.00
1941	318.66	+ 14.22
1951	361.08	+ 13.31
1961	439.24	+ 21.51
1971	548.15	+ 24.80
1981	683.32	+ 24.66
1991	846.30	+ 23.85
2001	1027.01	+21.34

Over the past few decades, the average rate of population growth has roughly been about 2.5 per cent annually. The percentage of decadal variation was more than 24 per cent in 1961–71 and 1971–81. But, during 1981–91, it has become 23.85 per cent and further down in 1991–2001 to 21.34 per cent. To provide food, clothing, shelter, education, employment and health care to this colossal mass of humanity would be a Herculean task. The impact of such a growing population on the environment is also a cause of great concern to all of us and all efforts should be taken for the harmonious relationship between population and environment.

The growth of population in India has created several environmental problems, some of which are:

(i) overall reduction in agricultural land,
(ii) lower productivity,
(iii) deforestation,
(iv) soil erosion, land slides, and expansion of wasteland,
(v) problem of drinking water,
(vi) malnutrition,
(vii) gradual change in climatic conditions,
(viii) increasing frequency of floods and droughts,
(ix) increasing level of pollution, and
(x) adverse effects on human health and quality of life.

In 1901, only 5.81 per cent population was living in cities, which increased to 18.92 per cent in 1951 and 25.63 per cent in 1971. In 1981, the percentage of urban population was 26.91, which had become 25.7 per cent in 1991 and 27.78 per cent in 2001. The number of million cities in India in 2001 was 35. The process of migration from rural to urban areas has created several socio-economic problems. The overall quality of urban environment has deteriorated over the years and now has reached a point where immediate attention is needed not only by the government but people also. The situation in metropolises like Mumbai, Kolkata, Delhi, Chennai, Bangalore, Hyderabad, Ahmedabad, Kanpur, etc., is becoming worse year after year, where even minimum shelter for a large section of population is not available. The total population of these metros accounts for nearly one-third of total urban population and 10.5 per cent of India's total population.

Due to uncontrolled urbanisation in India, environmental degradation has been occurring very rapidly and causing shortages of housing, worsening water quality, excessive air pollution, noise, dust and heat, and the problems of disposal of solid wastes and hazardous wastes. Urbanisation, associated with industrialisation, has created several environmental problems such as:

(i) growth of slums;
(ii) industrial pollution (air, water, soil);

(iii) air and noise pollution by vehicular traffic;
(iv) problem of solid waste disposal;
(v) sub-standard and unhygienic residential conditions;
(vi) problem of safe drinking water;
(vii) destruction of natural sites, vegetation, etc.; and
(viii) problem of waste water, etc.

Environment polices for cities in 1990s outline some basic principles that can be applicable to India. These are:

1. There is a need to intensify cooperation between different levels of government—local, regional and national—and each be made responsible to protect the urban environment.
2. Cooperation between the public and private sector must grow in designing solutions to environmental problems.
3. Minimum environmental standards must be established and enforced.
4. The local community must take part devising and implementing environmental policies. Local initiatives stimulate greater awareness of local problems and potential.
5. Prices must incorporate environmental and social costs. Local taxes and user fees are most appropriate for demand management and lead to resource conservation.
6. Synergies that can contribute to environmental protection must be identified. In a synergetic relationship, mutually enhancing activities are more productive when they are coordinated.
7. Short-term decisions must be taken within the framework of long-term strategies. Environmental strategy based urban planning can provide a long-term strategy within which short-term decisions should be taken.

There is also need to develop the environment-friendly cities. With systematic efforts by the government and communities, including NGOs, each city can be developed as a model environmental city.

Tasks Ahead

We have discussed above some measures for the improvement of urban environment. Apart from this, there is also need to plan strategy to limit population growth as well as for a balanced friendly ecosystem. Such measures are urgently required for countries having larger populations and also for all the developing countries. Some measures or action programmes could be as follows:

(i) Fertility reduction programmes should be intensified to constrain population growth.
(ii) Population welfare programmes like child care, immunisation, malnutrition control, safe drinking water, etc., and also programmes to relieve poverty, hunger and to raise socio-economic status of women should be implemented properly.
(iii) Rural development programmes related to massive employment in rural non-agricultural sectors, to conserve rivers, ponds, forests and grazing land in villages, to enhance the quality of life in rural areas by innovative technology for making cheap but durable housing, sanitation, drainage, water supply systems, etc.
(iv) Agriculture improvement programmes should be started along with horticulture, fishing, dairy and poultry.
(v) Knowledge of land-soil water management should be given to villagers so that they could adopt soil conservation measures and use available water resources.
(vi) The use of alternative sources of power such as bio-gas as well as alternative technology for bio-fertilisers should be popularised so that quality of environment could be improved.
(vii) Widespread social forestry and agro-forestry programmes should be implemented.
(viii) Conservation of forests and massive afforestation programmes should be part and parcel of a national programme.
(ix) For urban areas, apart from suggested measures, new urban land use policy, new modes of urban transport and rapid transit systems and urban slum development programmes

should be formulated for the improvement of quality of life in cities.

(x) Strict pollution control measures should be adopted without any political interference.

The above suggested steps are preliminary and achievable, if there is a political will, economic discipline, distributive justice and true commitment to ensure planning for balanced ecosystem.

Reference

Odum, Eugene P., *Fundamentals of Ecology*, 1971, p.162.

14

Environmental Hazards and Their Management

Natural hazards originate in highly dynamic processes whose elemental nature expresses itself in their unpredictable and uncontrollable character. Those natural hazards that cause losses of human life and great material damage to economics are called disasters and catastrophes. Two aspects of such natural hazards are noted: their potential for disaster and their actual catastrophic results. Not all potentially dangerous hazards are disastrous. In contrast to their potential danger which remains the same, the intensity of disasters is governed by five major factors:

1. Historical and social conditions, as well as the level of economic development of a country or a region, which affect their vulnerability.
2. The scope of the damage that natural hazards may cause derives from land use forms. It matters who and what are exposed to natural hazards.
3. The scope of disasters through natural hazards depends on a region's geographic position within a sequence of processes.
4. Natural hazards are often not so dangerous when taken separately but cause predicaments when combined with other processes.
5. In most cases the scope of disasters produced by natural hazards is governed not by a single characteristic but by the scale, duration and intensity of corresponding phenomena. While natural hazards are always potentially disastrous,

actually resulting damage will vary in accordance with social, economic and geographic conditions.

In studying the problem of natural hazards, the relevance of modern scientific and technical advances as well as the growing influence of human activities on natural environment should be recognised. These activities perturb and modify nature's original dynamic state. As a result, the self-regulating properties of natural processes either vanish or are weakened. It would be logical that having disturbed the environment's natural state, man should also assume corresponding regulation functions.

The scientific interest in environmental hazards is of fairly recent origin. Attempts to defend individual buildings against earthquakes date back at least 2,000 years and a concern for risk can be traced back to the earliest recorded times. But systematic research into natural hazards really began with the work of Gilbest White (1936, 1945), who was the first to recognise that engineering schemes were not the only way to tackle flood problems in the USA. Over next 20–30 years, the activities of White and his colleagues were extended to other geophysical hazards. As a result, hazard-related research became established in a wide variety of academic disciplines ranging from geology to sociology, not only in USA but also in Europe, Russia, Japan, China, etc.

The pace accelerated in the 1970s for several reasons. First, extreme natural events suddenly became more prominent. In large part, this was due to the climatic fluctuations of that decade which produced the Sahelian drought, the failure of the Peruvian anchovy harvest, the 1975–76 drought in northwest Europe and the severe North American winters of 1976–77 and 1977–78. These events exposed the vulnerability of many countries, including advanced ones, to climatic variability. Second, neo-catastrophism became fashionable amongst physical geographers, who also saw the need to make their work more relevant to human affairs and natural hazards became a strong focus for the study of the relationships between nature and society. Third, there was growing belief that some of the apparent inability to cope with hazards lay in differences between the 'real' world and how it was viewed in practice by managers and decision-makers. In turn, this gave momentum to the research on environmental hazard perception already begun by some human geographers.

Thus, the 1970s saw the publications of several important books. These works represent a plateau of achievements which, in the 1980s, was modified by three further changes in perspective. First, more emphasis was given to the relationships between underdevelopment and vulnerability to natural hazards in the Third World, in particular the extent to which economic dependency exacerbates the effects of geophysical events. Second, following important signals of technological hazards in the 1970s, such as the Flixborough (UK) explosion in 1974, the release of dioxin at Seveso (Italy) in 1976 and the nuclear incident at Three Mile Island (USA) in 1979 so-called 'man-made' hazards began to receive more attention. The year 1984 was a clear turning point with several major industrial accidents, including the release of methyl isocyanate at Bhopal (India) which alone has caused as many as 6,000 deaths. Third, the multidisciplinary nature of hazard research widened to the point where earlier distinctions between 'natural' and 'man-made' hazards were ever more difficult to sustain.

In terms of policy development, an international programme to reduce the losses from natural hazards—first proposed in 1984—was finally adopted in December 1989 by the United Nations General Assembly proclaiming the 1990s as the International Decade for Natural Disaster Reduction (IDNDR). Its objective was to reduce through concerted international action, especially in developing countries, the loss of life, property damage, and social and economic disruption caused by natural disasters, such as earthquakes, wind storms, tsunamis, floods, landslides, volcanic eruptions, wildfires, grasshopper and locust infestations, droughts and desertification, and other calamities of natural origin.

The terms 'hazard', 'risk' and 'disaster' are sometimes taken as synonymous but have some difference in nature and characteristics. *Hazard* is best viewed as a naturally, occurring or human induced process or event with the potential to create loss. *Risk* is the actual exposure of something of human value to a hazard and is often regarded as the combination of probability and loss. On the other hand, when large members of people exposed to a hazard are killed, injured or damaged in some way, the event is termed a *disaster.*

Environmental Hazards

Geographers have defined environmental hazards in terms of physical environment or in other words considered them as natural hazards. Accordingly, environmental hazards are those elements of the physical environment, which are harmful to man and caused by forces extraneous to him. But most of the environmental hazards have both natural and human components. For example, flood problems may be exacerbated by fluctuations in climate, such as increased storm frequency, and by human activities, such as land drainage and deforestation. The loss of life caused by a tropical cyclone can by greatly reduced by means of a warning message derived from monitoring by satellites and weather radar. Traditionally, natural hazards have been viewed in an ecological framework (see Figure 14.1). This distinguishes between natural events and their interpretation as natural hazards (or resource). Since the earth is a highly dynamic planet, most natural events show a wide range of variations through time in the use of energy and materials for environmental processes. The outer limits of this behaviour we call 'extremes' and certain statistical measures,

Figure 14.1

Environmental Hazards

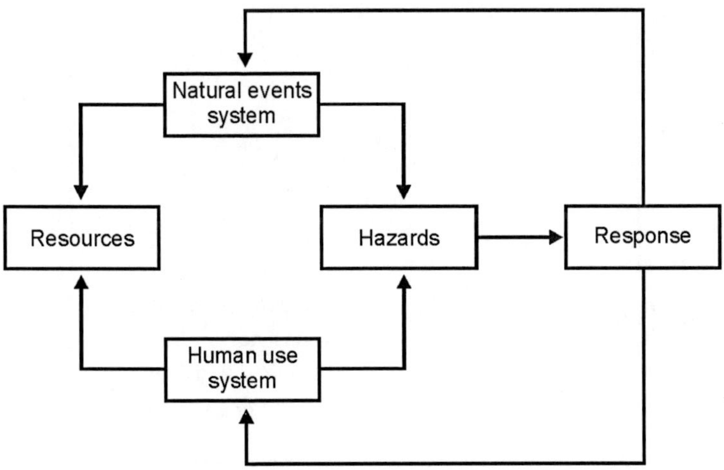

notably magnitude-frequency relationships, are used to describe such extremes. But extreme natural events are not considered disaster unless they cause large-scale damage to humans.

The term 'environmental hazard' has the advantage of both natural and human dimensions. It also implies a spectrum of hazard types. Figure 14.2 shows that hazard range from 'natural' (geophysical) events, through 'technological' (man-made) event to 'social' (human behaviour) events. Some attempts can also be made to scale the range of hazards according to whether the impacts are intense and local, or diffused and widespread within society. The extent to which hazards are voluntary or involuntary is particularly important. The degree of individual human responsibility for disaster increases greatly from the essentially accidental geophysical hazards (earthquake, tsunami) to the large self-induced social hazards (smoking, mountaineering).

Figure 14.2

General Spectrum of Environmental Hazards from Geophysical Events to Human Activities

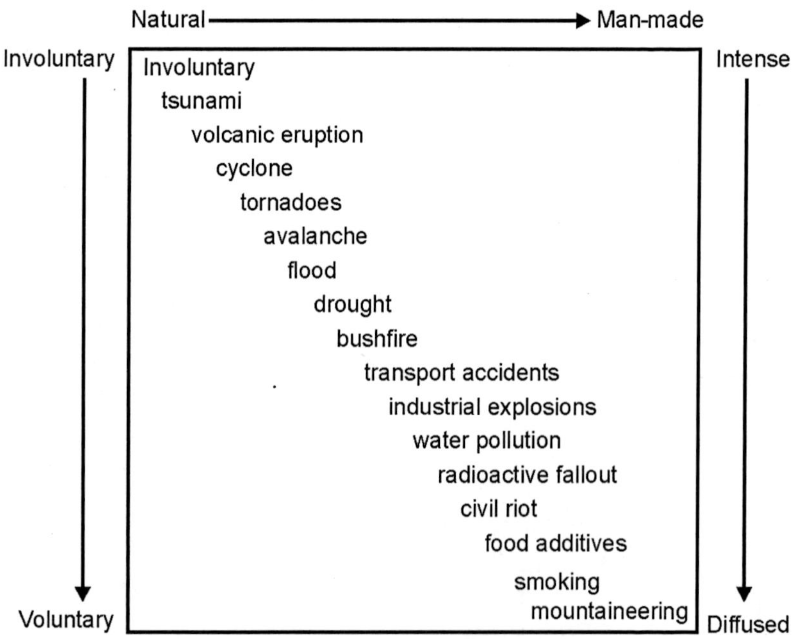

Figure 14.2 represents a more wide definition of environmental hazard, e.g., "the threat potential posed to man or nature by events originating in, or transmitted by, the natural or built environment" (Kates, 1978). The common features of the environmental hazards/disasters are:

(i) The origin of the damaging process or event is clear and produces characteristic of threats to human life or well-being.
(ii) The warning time is normally short.
(iii) Most of the direct losses, whether to life or property, are suffered fairly shortly after the event, i.e., within days or weeks.
(iv) The exposure to hazard, or risk, is largely involuntary, normally due to the location of people in a hazardous area.
(v) The resulting disaster occurs with an intensity that justifies an emergency response.

Thus, the working definition of environmental hazard is: "extreme geophysical events, biological processes and major technological accidents, characterised by concentrated releases of energy or materials, which pose a largely unexpected threat to human life and can cause significant damage goods and the environment."

Types of Environmental Hazards and Disasters

There are several types of environmental hazards. But, a general classification is necessary for proper understanding and analysis. While classifying the environmental hazards and/or disasters, the following points should be taken into consideration:

(i) areal extent of damage zone,
(ii) intensity of impact at a point,
(iii) duration of impact at a point,
(iv) rate of onset of the event, and
(v) predictability of the event.

Hewitt and Burton (1971) have listed potentially hazardous environmental agencies or processes as given in Table 14.1.

Hohenemser (1983) has viewed technological hazards as a sequence of events leading from human needs and wants to the selection of a particular technology, though to harmful consequences.

A schematic illustration of the causal chain of hazard development with an example of drought disaster has been depicted in Figure 14.3. Here, the top line indicates seven stages of hazard development. The stages are linked by causal pathways denoted by triangles. Six control stages are linked to pathways between hazard states by vertical arrows. Each is described generally as well as by specific control actions designed to eliminate or reduce the evolving hazard.

Hazard can also be divided according to the damage caused by them. Although, such division is not effective in all parts of the world because of the difference in development, technology, etc. For example, an ordinary disaster may include:

— at least 100 people dead, or
— at least 100 people injured, or
— at least US $ 1 million damage.

Figure 14.3
Causal Chain of Hazard Development

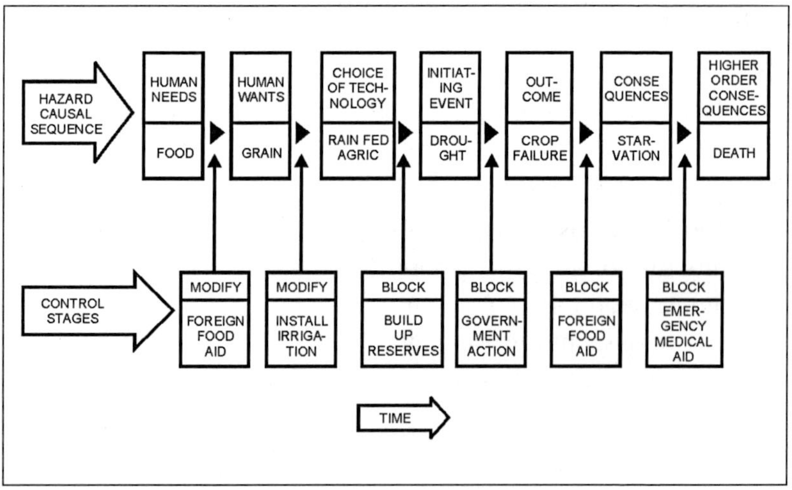

Table 14.1

Potentially Hazardous Environmental Agencies or Processes

1. *Atmospheric*

Single element	*Compound hazards*
Rain	Rain and wind storms
Freezing rain('glaze')	'Glaze' storms
Hail	Thunderstorms
Snow	Tornadic storms and tornadoes
Wind	Hurricanes
Lightning	Blizzards
Temperature: 'heat wave', 'cold spell', frost	'White-out'
Fog	Drought

2. *Hydrologic*

Flooding: riverine (rain, snowmelt, natural damburst floods)
Lake and sea-shore wave action
Waterlogging
Sea-ice and icebergs
Run-off drought
Glacier advance

3. *Geologic*

Mass-movements: landslides, avalanches, mudflows, subsidence, etc.
Erosion (foundations, soils, etc.)
Earthquakes
Volcanic eruptions
Shifting sands

4. *Biologic*

Severe epidemics in humans
Sever epidemic in plants
Sever epidemic in domestic and wild animals
Animal and plant invasions (e.g., locusts)
Forest and grassland fires

5. *Technologic*

Transport accidents
Industrial explosions and fires
Accidental releases of toxic gas
Nuclear power plant failures
Failures of public buildings or other structures
Germ or nuclear warfare

But such occurrences have great spatial differences in less developed countries and more developed countries, therefore, classification based on losses in not effective.

The polygenic character of natural hazards of the world derives from its diversity of natural conditions. Potentially, disastrous natural phenomena are also closely associated with certain regions and geographic zones. Such phenomena often develop at the interface of physically different natural environments: sea and land, atmosphere and ionosphere, etc. In the world most common and disastrous hazards are: volcanoes, earthquakes, landslides, floods drought and famines, thunderstorms, tornadoes and tropical cyclones.

Volcanoes

The eruption of volcanoes is one of the most dreadful natural phenomena. The molten rocks, boiling lava and hot gases penetrate out of that part of the surface which is comparatively weak. When this material comes out of these spots, an explosion takes place, with thunderous noise, molten rocks, lava and gaseous clouds rise up towards the sky. All this results in a threat to the environment as well as to life.

Volcanism is not common throughout the world but is limited in the Circum-Pacific belt extending from Andes in South America, through the Cascades and the Aleutians into Japan and then south into the East Indies and New Zealand. There is also an important belt in the Mediterranean, which includes the active volcanoes of Italy.

A volcanic explosion poses a great threat to environment. On an average, a volcano, in an explosion, ejects 150 cubic km of material. This material is of three types: solid material in the form of small and large pieces of rocks, ash particles and also pieces or volcanic bombs which shoot into the sky. The liquid material called *magma* comes out of volcano—also known as *lava*. A huge amount of gaseous material comes out of volcano—steam, carbon dioxide, sulphur in vapour form produce poisonous gases when they combine with other gases. Gases like chlorine, fluorine, etc., are also emitted in large quantities.

The eruptions of volcanoes and lava flows are environmental hazards of the severest type, often taking a heavy toll of plants and animal life. Loss occurs from sweeping clouds of incandescent gases and suspended matter that descend the volcano slopes like great avalanches. When Mont Pele burst in 1902 in the West Indies, its burning material ejected out of it moved in the form of a cloud over the city called St. Piere and only two people out of total 30,000 population survived. The explosions of Etna and Vesuvious in Italy destroyed many beautiful cities. In 1943, when Paracutin, 320 km west of Mexico city, exploded, it created a mountain 167 metre high. In 1991, volcanic explosions in Japan and Philippines have damaged a lot of property and caused loss of human life and also polluted the environment. Similarly, volcanic activity in the Andaman & Nicobar Islands was responsible for the environmental degradation in this area.

Earthquakes

Earthquakes are tremors, which are produced by the passage of vibratory waves through the rocks of the earth. The devastation caused by earthquakes to the human life and property, as well as to the ecosystem, is very high. The earthquake belt of the world is also associated with the volcanic belt, but several other regions too have occurrence of earthquakes. Wherever earthquakes occur, apart from loss of life and property, landslides, change in course of rivers, loss of vegetation and animal life, destruction of dams and development of faults and fractures in earth's surface are common and all this leads to environmental degradation. The list of devastating earthquakes in the world is too long to be enumerated here. In India, the occurrence of earthquakes in areas of Jammu & Kashmir, Himachal Pradesh, Uttar Pradesh, Bihar, West Bengal, Punjab, Delhi, Haryana, and Maharashtra north-eastern states are common. The most recent earthquake occurred on 26 January, 2001 in Gujarat and on 28 March 1999 in the Kashmir in which heavy loss of property and life was reported, apart from the whole regional ecosystem being disturbed. Major earthquakes that have occurred in the world are as follows:

Date	*Place*	*People died*
18th April, 1920	San Francisco	More than 3,000
1st Sept., 1923	Tokyo	1.42 lakh people
26th July 1963	Mecedonia	1,000
31st May, 1970	Peru	66,000
28th Dec.,1972	Nicaragua	10,000
19th Sept., 1985	Mexico	10,000
7th Dec., 1988	N.W. Armenia	25,000
21st June, 1990	Gilan city of Iran	40,000
30th Sept., 1993	Western & Southern India	10,000
17th Jan., 1995	Kobe (Japan)	6,430
27th May, 1995	Sakhalin Island	1,989
May, 1997	East Iran	1,600
30th May, 1998	North Afghanistan	4,000
17th Aug., 1999	Turkey	4,000
21st Sept., 1999	Taiwan	2,500
26th Jan., 2001	Gujarat(India)	20,000
21st May 2003	Algeria	2,000
26th Dec., 2003	Bap city (Iran)	41,000
25th Feb., 2004	Morocco	400

In fact, earthquakes are most devastating and disastrous among natural hazards.

Floods, Droughts and Famines

Floods and droughts both are natural hazards and are related with rainfall. If there is heavy rain in the catchment area of a river, flood occurs. On the other hand, if there is less rain or failure of monsoon, drought occurs. In both the situations their impact on environment creates problems and also affects the regional ecosystem.

Flood is a hydrological phenomenon that can occur in any drainage basin. Its magnitude and frequency varies primarily on the precipitation conditions in the catchment area. Most of the river valleys are subjected to varying degrees of flood incidences, either creating minor damages or devastating havoc at times. Such calamities of floods pose a serious problem to the whole environmental set-up of a region, bringing about some phenomenal changes in the physical environment ushering in accelerated erosional transportional and depositional activities of the river with the consequent effects on soil. The devastating effects of flood can be seen on vegetation, agricultural, industry settlement, sometimes causing loss of human and animal lives and of material also.

The main cause of flood is excessive rainfall but its occurrence is also due to earthquakes, deforestation in the upper catchment areas, failure of dams, excessive snow melting, etc. The beneficial effects of floods have been established beyond doubt in the form of fertilisation of soil. But, its devastating effects far exceed the benefits.

In India, the problem of flood is of gigantic magnitude, because of: typical seasonal rainfall, and erratic nature of tropical cyclones and rainstorms. Besides in last few years, most parts of the country, including metropolitan cities like Mumbai, have been badly affected by floods—the reason being unplanned development. In August-September, 2008, devastating flood has been occured in parts of Bihar. This was due to excess water release in rivers originated in Nepal.

The following are the areas which may be termed as flood-prone:

(a) Upper catchment and piedmont zone of the Himalayan river basins.
(b) Ganga river valley—upper, middle and lower deltaic zone.
(c) Brahmaputra valley and its tributaries.
(d) Deltaic region of Mahanadi, Godavari, Cauvery and Krishna rivers.
(e) Narmada and Tapti basins.

The common measures of the flood control could be:

(i) Control of erosion in upper catchment areas.
(ii) Construction of dams and reservoirs at suitable sites.
(iii) Prevention of overflow in channels.
(iv) Construction of multiple channels for distribution of the excess river water.
(v) Development of shelters in flood-prone areas.

The problem of flood should be taken up with more intensive preventive measures, in order to avoid environmental problems.

Drought is also a natural hazard, which has its impact on the regional ecosystems. The areas which get less rainfall, i.e., amount of rain is too less to support agricultural production or even grasslands, are usually called drought-prone areas. When droughts become severe, they cause *famines*, which are devastating for man, animals and also environment. Generally, droughts occur when rainfall is less than 25 to 50 per cent of the average. If it is below 50 per cent of the average, famines occur. In 1899, India faced the severest famine known as *Chappania Akal.* Another devastating famine occurred in 1977. In both these several lakhs of people lost their lives. In 1992–93, Ethiopia also suffered from the grips of a bad famine, in which more than 30 lakh people died. Famine conditions are common in Rajasthan, Gujarat, Haryana, Madhya Pradesh, Orrisa and Andhra Pradesh. In Rajasthan, it has become a regular phenomenon.

Thunderstorms, Tornadoes, Tropical Cyclones and Landslides

Due to abrupt changes in atmospheric conditions thunderstorms, lightning, hailstorms, tornadoes and cyclonic storms occur. A thunderstorm is an intense local storm caused by strong conventional activity. Thunder and lighting normally accompany the storm, and rainfall is heavy, sometimes due to cloudburst also. Violent surface wind may occur at the onset of the storm. The speed of the rising air in a thunderstorm is about 112 km per hour. *Lightning* is also one of the environmental hazards that results annually in the death of many persons and livestock and causes forest and building fires. One of the most destructive products of the thunderstorm is the hailstorm. Hail consists of rounded lumps of ice ranging from 0.5 to 5 cm in diameter. Large hailstorms are an environmental hazard as they can be extremely destructive to crops and flimsy buildings.

A *tornado* is a small, intense cyclone, in which the air spirals at a tremendous speed. Associated with frontal airmass interactions in mid-latitude, tornadoes are the most violent storms. These storms are most frequent and violent in the United States although occurring in Australia and in other places of mid-latitudes. A tornado moves across the country with a wind speed of upto 400 km per hour in a funnel-like shape. The end of the funnel cloud may alternately sweep the ground, causing complete destruction of anything in its path.

Tropical cyclones are the most powerful and destructive types of cyclonic storms, which are also known as *hurricane* or *typhoon.* These storms develop over oceans in latitude 8°to 15° N and S, but not close to the equator. The tropical cyclone is an almost circular storm having low pressure in which winds spiral at high speed accompanied by heavy rainfall. The environmental impact of tropical cyclones lies in the form of destruction of life, infrastructures as well as forest and animals. In islands and coastal regions, large-scale destruction of cities and their inhabitants has been reported on several occasions. Bangladesh is a country which has suffered a lot by heavy destruction caused by tropical cyclones—the death toll by tropical cyclones in Bangladesh is almost unbelievable. In 1822, the toll was 40,000 while in 1876 and 1897 it was 1 lakh and 1.75 lakh respectively. In 1970, 3 lakh lives were lost and in 1991 the death toll was 5 lakh. These storms are also responsible for change in atmospheric and environmental conditions.

Landslide is another event which may cause disruption in the general ecosystem. It is a natural hazard and means any downslope movement of mass of regolith or bedrock under the influence of gravity. It is a rapid sliding of large rock masses beginning their descent as unit of blocks, without internal flowage. When the landslide is in the form of huge bocks, it is known as *rock avalanche.* In Switzerland, Norway or Canadian Rockies, villages built on the floors of steep-sided valleys have been destroyed by the sliding of rock masses. In India, reports of rockslide often appear in the news in hilly states like Jammu & Kashmir, Himachal Pradesh and north-eastern states. The environmental hazard of rockslides is extremely high for towns located on the floors of glacial troughs and at heads of fiords. The blockage of roads by landslide is a common phenomenon but sometimes it also blocks the course of the small rivers, thus, causing much damage.

Natural Hazards and Their Impact on Environmental in India

In every part of the world, several natural phenomena, such as volcanic eruptions, earthquakes, cyclones, landslides, floods, droughts and famines, have had their impact on environment.

Sometimes their impact is sudden and normal but often it is adverse not only to man but also to the environment and therefore are termed as natural hazards, environmental hazards, environmental stresses, environmental disasters, extreme events, etc. These natural hazards are common in various parts of India.

Since India is not located in the volcanic zone, volcanic eruption is not common. But, in November-December 1991, at Andaman Island, a dormant volcano became active and caused damage to vegetation, marine life and also polluted the atmosphere. The incidence of landslide is common in hilly areas of Jammu & Kashmir, Himachal Pradesh and Uttrakhand. Similarly, coastal regions are susceptible to cyclonic storms, which cause massive destruction of life and property as well as to vegetation. The coastal regions of Orissa, West Bengal, Andhra Pradesh and Tamil Nadu often face the problem of cyclones.

In India, earthquake-affected areas include Jammu & Kashmir, Himachal Pradesh, West Bengal and the north-eastern states. These are located in the foothills of Himalayas, while there are other states, such as Punjab, Haryana, Delhi, Rajasthan and Maharashtra, which are also affected by earthquakes. The impact of earthquakes on environment is much more in the form of destruction of flora and fauna, occurrence of rifts and cracks in land, change in course of rivers, cracks in dams and a great loss of man and property, depending upon the intensity of the quake. On 11 October 1937, an earthquake claimed the lives of over 3 lakh people. Other important earthquakes have occurred on 30 May 1885 in Kashmir, 12 June 1887 at Shillong, 15 June 1870 in Kacch, 4 April 1905 in Kangra Valley, 5 January 1934 in Bihar, 15 August 1995 in Assam, 11 December 1967 at Koyana (Maharashtra), 21 August 1988 at Darbhanga (Bihar), 20 October 1991 in Uttar Kashi, 30 September 1993 and 26 January, 2001 in Maharashtra, etc.

In India, occurrence of floods during the monsoon season is very common. Floods are very common in states like Bihar, Uttar Pradesh, Uttrakhand, Andhra Pradesh, West Bengal, Assam, Gujarat, Tamil Nadu, Haryana, and Madhya Pradesh. In India, the total area affected by floods is about 1.37 million hectares, while the total affected crop area 157.6 million hectares. Effective flood control measures, such as utilisation of flood water by construction of small dams, afforestation in catchment area and construction of channels, etc., can minimise its hazardous effect.

While a good monsoon becomes a cause of floods in India, the failure of monsoon is responsible for drought resulting in famines. Droughts in turn become cause of environmental degradation. The famine that occurred in year 1899, known as *Chappania Akkal*, is well known in the history of India, because of its widespread impact and death of several lakhs of people. A similar famine occurred in 1917. In 1986–87, about two-third of the country was under the grip of famine. The worst hit area is Rajasthan, where famine often occurs due to desertic conditions and failure of monsoon. During the last 20 years, Rajasthan has had famine in 16 years. The development of irrigation facilities and proper water management can help solve this problem.

Management of Environmental Hazards

Adjustment to Hazard

Most environmental hazards, which pose a recognisable threat to human populations, produce some response designed to reduce the level of risk. Theoretically, the actions taken can attempt to manage the full range of hazards which affect an area but the information and resources necessary for such an approach are rarely available. At the other extreme, sometimes, nothing can be done but simply accept recurrent losses. Within this range, all intentional hazard adjustments can be organised into three groups. These are:

- *Modifying the loss burden*: The most limited responses seek to modify losses by spreading the financial burden as widely as possible beyond the immediate victims through relief and insurance programmes. These schemes are essentially loss-sharing, rather than loss-reducing, measures. But, in some circumstances, both disaster aid and insurance can also be used to a limited extent to encourage future loss reduction.
- *Modifying hazard event*: This involve reducing losses by adjusting damaging events to people. Most natural hazards are insufficiently well understood, or manageable, to be physically suppressed at source through environmental control

engineering. On the other hand, hazard-resistant design and emergency measures can be applied to safeguard lives and property from selected phenomena in certain high-risk settings.

- *Modifying human vulnerability*: This is the largest group of responses. It includes all measures designed to reduce hazard losses by adjusting people to damaging events. This is achieved mainly through preparedness programmes aimed at changing human behaviour, through the installation of forecasting and warning schemes and through long-term planning to encourage more appropriate land use allocations.

Hazard Resistance

This requires hazard-resistant design and retrofitting. In most countries, public facilities such as dams, bridges and pipelines are likely to be hazard-resistant because they would have been designed by professional engineers. The same is true for structures such as nuclear power plants and chemical factories. However, design protection can never be total. Building failure will be greatest in data-poor, hazard-prone environments, where many structure—notably private residences, older and smaller buildings—would have been constructed without any thought to hazard impact. The two most common natural hazards considered in building codes are earthquakes and windstorms, although the adjustment is also applicable to other hazards such as floods.

In the case of earthquakes, hazard resistance begins with geotechnical engineers who apply the principles of rock and soil mechanics to the safe design of earth-supported structures. Other things being equal, buildings on solid rock are less likely to suffer damage than those built on clays or softer foundations. Microzonation maps of the substrate can be prepared which suggest appropriate local variations in building strength and ensure that major buildings are not located over faults or areas of unconsolidated material. Local building codes tend to undergo continuous improvement and the adoption and enforcement of such improvements is the primary means of increasing the hazard resistance of structures. In fact, proper hazard-resistant design could make structural property safe and its insurance will also provide damage loss, if any.

Retrofitting is the act of modifying an existing building to protect it, or its contents, from a damaging event. Many individual measures can be taken. In the case of earthquakes, for example, brick chimneys could be reinforced and braced to structural elements to prevent collapse into a living area. Unreinforced masonry walls could be strengthened and tied to adequate footings while closets and heavy furniture could be strapped to the walls whenever they constitute a danger or contain valuable items. To protect against floods, measures include making the walls water-tight and fitting flood-resistant doors and windows.

Community Preparedness

Disaster preparedness may be defined as "the pre-arranged emergency measures which are to be taken to minimise the loss of life and property damage following the onset of disaster". It involves the detailed planning and testing of prompt and efficient responses by both individuals and groups, once a hazard is either forecast or has actually occurred. Measures include the promotion of public education and awareness programmes, the development of evacuation plans (often in response to a warning message), the local supply of medical aid and the preparation of emergency food and shelter for evacuees.

Long-term community preparedness programmes have proved very successful in reducing deaths from hazards in many developed countries. They are particularly important adjustment for hazards, like earthquakes and volcanic eruptions, for which reliable short-term forecasts and warnings are not routinely available. Within individual countries, the degree of preparedness usually depends on the arrangements which have been made for civil defence.

For a successful operation of any disaster plan, public understanding and cooperation are vital elements. Most of the people are not prepared to handle the consequences of a damaging event. Simple preparedness at domestic level such as a battery-operated radio, a working torch and a first aid box will help a family during any emergency.

Good educational programmes are a key element in ensuring that widespread public support is available for hazard mitigation, as well as facilitating the transfer of disaster experience from one community to another. Ill-conceived awareness exercises may create panic and other counter-productive attitudes, so it is important that such programmes are well based in terms of social feasibility and an understanding of the probable economic consequences of the mitigation measures recommended to the public. Recent experience suggests that carefully prepared advice need to be distributed, both widely and often, to the public through the media and other channels. All preparedness advice should come from an authoritative government agency and should be locally endorsed by well-known local officials and public figures. Each measure should be accompanied by a brief explanation which justifies the recommended action and details on how it can best be implemented.

Forecasting and Warning

Forecasting and warning systems have become important in recent decades owing to scientific advances, for example, in weather forecasting, and to associated improvements in information and communication technology such as satellites. Most warnings of a future environmental hazard are based on forecasts. But some threats, notably earthquakes and droughts, are insufficiently understood for forecasts to be routinely issued and reliance for warning has to be placed on predictions instead. It is important to understand the difference between predictions, forecasts and warnings.

A *prediction* is largely based on statistical theory and uses the historical record of past events to estimate the future probability or recurrence of similar event. Since the results are expressed in terms of average probability, there is no precise indication of when any particular event may occur. By definition, predictions tend to be relatively long-term. For some hazards, such as earthquakes, it is not possible to specify the location or the magnitude of the event with much precision. A *forecast*, on the other hand, depends on the detection and evaluation of an individual event as it evolves through a sequence of environmental processes which are relatively well understood. This means that, depending on the ease with which such individual events can be monitored, it is often possible to specify very precisely the timing, location and magnitude of an impending event. Forecasts are based on scientifically observable facts and offer no prescription as to how people should respond. They tend to be short-term. Indeed, contrary to predictions, the limited lead time for issuing forecasts often restricts the effectiveness of warnings.

A *warning* is a message, which advises the public at risk about an impending hazard and also what steps should be taken to minimise losses. All warnings are based on either predictions or forecasts but for many agencies, such as those involved with national weather services, very few of the routine forecasts are

Figure 14.4

Hazard Forecasting and Warning System

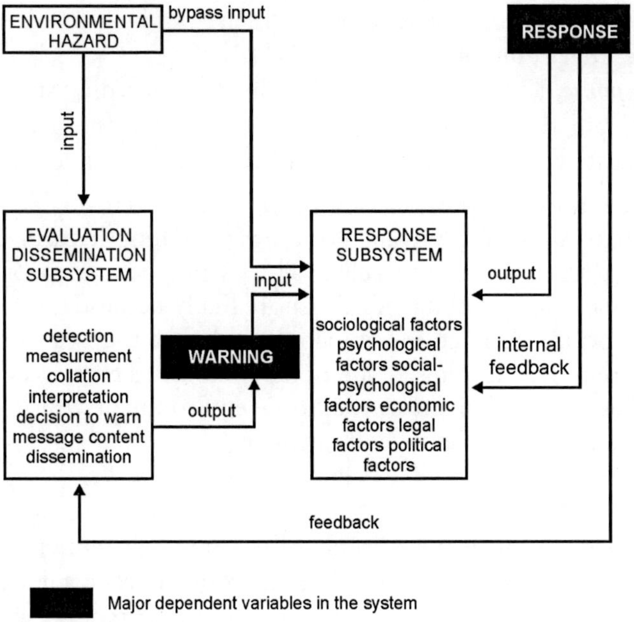

Source: White and Hass (1975).

followed by warnings. All warning systems consist of three interrelated stages of evaluation, dissemination and response.

- *Evaluation*: As shown in Figure 14.4, the evaluation stage includes several sub-steps from trained observers' first detecting an environmental change that could cause a threat, through to estimating the nature and scale of the hazard and finally deciding to issue a warning to an endangered community. The priority at this stage is to improve the accuracy of the forecast and to increase the lead time between issue of the warning and the onset of the hazardous event.
- *Dissemination*: The dissemination stage occurs when the warning message is transmitted from the specialist forecasters to the hazard-zone occupants. The message is likely to be formulated and conveyed by a third party through an intermediate medium, which may involve different communication methods, such as radio or television, and different personnel, such as the police or neighbours. Again, the stage contains several components, like the content of the message or the way in which it is conveyed, which are known to affect the eventual outcome.
- *Response*: The whole purpose of the two preceding stages is to activate the response stage, where loss-reducing actions are expected to be taken by the recipients of the warning message.

One important purpose of warning systems is to enable people to evacuate an area in advance of an approaching hazard. Often, there is a reluctance to evacuate. This is not because people are paralysed in the face of danger but more likely because the message fails to specify this action, or people feel that they can cope or because they fear looting of their empty house. There is a considerable natural attachment to the home environment which has to be overcome before people will evacuate.

Land Use Planning

Land use planning is a comprehensive approach, which include tools such as regional planning, zoning ordinances and sub-division regulations. It seeks to intervene in the process whereby hazard-prone land, initially held in low-intensity uses such as forestry or agriculture, is converted into higher intensity occupation. Such occupation increases unit land values and leads to greater potential losses. The main purpose is to guide new residential, commercial and industrial development away from identified hazard zone—but this is not always possible. Therefore, land use management has an additional role to play in lessening potential losses in areas already occupied. Land use planning is most useful in communities that are still growing and also have undeveloped land. Successful management techniques also depend on the availability of information with which to identify particularly hazardous locations. Indeed, the accurate delimitation of the hazard zone is crucial because the entire policy is based on the detailed recognition and community acceptance of different degrees of risk which in turn justify the implementation of selective development controls. Ideally, variations in risk should be identifiable down to the level of individual properties. For many hazards, such as cyclones and earthquakes, such precision is unattainable. The greatest accuracy is achieved with topographically controlled hazards like floods, landslides, avalanches and frost.

Once the high-risk areas have been identified, a number of management options are available for reducing losses. Public acquisition in hazard-prone land is the most direct measure available to local government and is one of the most effective long-term strategies. Once acquired the lands can be managed to protect public safety or to meet other community objectives, such as open space or low-density recreational facilities. But land acquisition is expensive and local authorities rarely have the resources for outright purchase.

In brief, environmental hazards are prevalent throughout the world and are the result of natural phenomena as well as human activities. There is thus need to prepare both short-term and long-term plans so that the loss can be minimised. A proper management system, broad community support and a strong political will is very necessary for the adjustment and control, whenever a natural hazard occurs.

15

Global Environmental Issues

During the 1960s and 1970s people began to think that human existence would be threatened unless they adjusted their relationship with their environment. Now, it has been realised that environmental problems are not limited to local, regional and national levels but there are also several global environmental problems. The scientific and technological revolution has given multiple facilities to mankind but at the same time it is responsible for depletion of resources, excessive use of fossil fuels, deforestation, desertification, loss of the fertility of soil, changes in atmospheric conditions resulting into serious problems like greenhouse effect, depletion of ozone layer, rise of world temperature, etc. All these and many other problems have a global dimension and the world community has already realised this fact that cooperative efforts can only, if not solve, at least, minimise these environmental problems. In this chapter, a brief introduction to certain global environmental problems has been given, which will help to think over global issues and to develop a public opinion regarding these issues.

Thinning of Ozone Layer

Ozone is a form of oxygen with three atoms instead of the normal two, and forms a fragile shield scattered in the stratosphere absorbing the sun's ultraviolet radiation, away from the earth's surface. It forms a 3 mm thick layer, but if it disappears or thins, all terrestrial life will be annihilated. Near the earth's surface, ozone is an increasingly troublesome pollutant, but safely up at a height of

20 to 30 km from the earth in the atmosphere, it is as important to life as oxygen itself.

The thinning of the ozone layer has generated global concern in the past decade. This is due to several chemical pollutants discharged by industries or produced through other chemical reactions. Chlorofluorocarbons (CFCs) are substances that are inert, inexpensive, stable, non-inflammable, non-toxic, and easy to store and produce. The population explosion is one of the causes of the increase of CFCs due to use of airconditioners, insulation, solvents and fire-killers. In 1986, three US agencies, viz., National Aeronautics and Space Administration (NASA), National Oceanic and Atmospheric Administration (NOAA) and Federal Aviation Administration (FAA) started reporting on ground-based and satellite-based data on ozone depletion in stratosphere by the use of chemicals like halon 1301, methyl, chloroform, carbon tetrachloride, solvents FC-113, aerosols, foams and refrigeration CFC-11, and air conditioning CFC-12 that are depleting the ozone layer 4, 5, 8, 12 and 26 per cent, respectively, on global scale, because of human activities. As far as CFC consumption is concerned, USA is on top consuming 29 per cent, followed by Russia and Europe 14 per cent, while in China and India it is merely 2 per cent. The total ozone amounts in the northern hemisphere, middle and high latitudes have decreased, and sudden and unexpected decreases in the amount of spring time Antarctic zone is confirmed on 15 to 20 km in the stratosphere ozone layer.

The increase in ozone layer depletion will invite the lethal ultraviolet rays from the sun which will cause cancer, eye damage, injure plants and marine life and even reduce our immunity to diseases. The UV-B damages the genetic material DNA and also causes skin cancer. The ultraviolet radiation even reduces the efficiency of the immune system, thus impairing the body's resistance. Experts are of the opinion that animal and plant world would face a grim challenge if radiations are not checked. Certain crops have been found vulnerable to such radiations. Under the impact of radiations proteins in plants, microbes and animals would be affected and such changes in spatial orientation would be critical in the formed part of the reaction centre of an enzyme.

Near cities where population concentration is more, ozone could be formed near the earth's surface having disastrous effects on human health, crops and ecosystems. It would also increase the incidence of smog in areas of industrial concentration. Ozone has had its effect on the earth's climate by adding to the greenhouse effect.

The world is now much serious about the most dramatic evidence of ozone depletion—a "hole" in the ozone layer, which appears every spring over the Antarctica. This hole has grown over the last ten years. Till now, no clear explanation has been found to explain this phenomenon, though scientific research in this field is in progress. But, the general opinion regarding the cause of this depletion is the CFCs. Since this is a serious problem, this becomes the duty of the developed countries, having both technology and infrastructure, to check the ozone depletion. Likewise since this is not a problem of individual nations, therefore, combined efforts can only check it.

The concern about ozone layer protection began by the late 1970s and early 1980s. In 1978, the United States banned the principal CFCs for most uses; in 1980 European Community countries placed a limit on production of CFCs and cut their use in aerosol products by 30 per cent. Sweden, Norway and Canada also placed limits on CFC use. In 1975, UNEP has also expressed serious concern about ozone depletion and also developed a World Plan of Action on the Ozone Layer. Some of the important international actions taken on ozone protection are listed below:

1977	UNEP Coordinating Committee on the Ozone Layer established.
1985	Vienna Convention for the protection of ozone layer.
1987	Montreal Protocol on substances that deplete the ozone layer.
1988	Entry into force of the Vienna Convention.
1989	Entry into force of the Montreal Protocol.
1990	Montreal Protocol amended in London. The amendment required signatory governments to regulate consumption and production of CFCs.
1991	Establishment of interim multilateral fund to provide financial assistance to developing countries in order to facilitate control measures.
May 1992	112 states have ratified the amendment to the Montreal Protocol.

Aug. 1992	Entry into force of the amendment.
Nov. 1992	At Copenhagen, parties agreed to accelerate, phase out schedule, to January 1996.

Ozone Protection Regime

Both the state and non-state actors played major roles in the regime-building process. To a large extent, their roles were determined by their position in the global economic system.

At the beginning of the problem definition process, there was scientific uncertainty about the gravity of the ozone depletion problem. The definition process reduced this uncertainty by accomplishing three major tasks: it confirmed that ozone depletion was occurring; it confirmed the connection between the use of CFCs and halons and the damage to the ozone layer; and it estimated the extent and rate of depletion.

The ecological epistemic community, which played a major role in defining the problem, was a transnational one. It was made up of atmospheric scientists and policy makers, including officials of UNEP, the US Environmental Protection Agency, and the State Department's Bureau of Oceans and International Environmental and Scientific Affairs (OES). The epistemic community used a variety of channels to affect the decision-making process. In the process of drafting documents and reports, gathering data, organising scientific panels and lobbying delegates, it was able to substantially define the issue and modify the agenda for the negotiations. Significant components of the epistemic community participated in UNEP's Coordinating Committee on the Ozone Layer, in which they were able to share information.

The bargaining process over the ozone protection regime actually began before the issue was clearly defined. In the beginning, the Third World countries played a minor role in the negotiations, primarily because they did not have clearly defined interests. However, nations with economic interests, as well as industries involved in CFC production, did have a clear understanding of the interests as stake.

Nations brought dissimilar viewpoints to the negotiations. Some favoured a global ban on CFCs 11 and 12 in aerosols (except for essential purposes) and a limit on CFC emissions in non-aerosol uses. The United States, which accounted for 30 per cent of worldwide production, was prepared to be the lead state. However, other nations were reluctant to undertake such sweeping actions until they felt the scientific evidence against CFCs was compelling. The latter group would have been satisfied with a framework convention.

The European Community accounted for 45 per cent of world CFC output, and by the mid-1980s it was exporting a third of its production to developing states. Because of the major producer states' interests in continuing production, the EC member-states constituted a veto coalition. Since participation of these states was necessary for a strong ozone protection regime, their cooperation was important. The large developing states, such as India, China, Indonesia, Brazil and Mexico, also appear to have had some potential as a veto coalition. Their bargaining leverage was based not on their consumption of CFCs but on their potential as producer and consumers of CFCs. These countries were already producing on a small scale; although they were providing less than 5 per cent of the world's CFC production, their production was increasing by 7 to 10 per cent each year. However, in the negotiations on the Vienna Convention, these states did not take advantage of their veto power. At this stage, no Third World country played an active role.

The Toronto Group took the lead, pushing for simultaneous negotiation of a framework convention and associated protocols that would obligate states to reduce CFC use. However, the veto coalition would not support the negotiation of regulatory protocols, taking the position that the state of scientific knowledge was insufficient to support such a protocol. The lead state remained on the offensive, proposing a worldwide ban on non-essential uses of CFCs in spray cans in 1983. But, the veto coalition offered a counter-proposal: total CFC production by 30 per cent beyond the anticipated 1985 level.

Over a three-year period, there were eight negotiating sessions, but the groups failed to achieve consensus on the ozone issue. In 1985, after three years of negotiation, the United States began to define ozone depletion as an urgent issue, but still the veto coalition would not yield. As a result, the parties had to be satisfied with the Vienna Convention, which was a framework convention.

The Montreal Protocol began the process of strengthening the ozone layer protection regime. Because of new findings, most nations accepted the idea of some controls. However, the extent and the timing of the controls were contentious issues. Lead states argued for a freeze, followed by a gradual reduction in the production of ozone-depleting substances over a period of 10 to 14 years, whereas the veto coalition wanted a production cap. The Toronto Group still pressed for a cut, but as a compromise its members suggested a 50 per cent cut. As late as April 1987, the EC insisted it would agree to a reduction of no more than 20 per cent, but at the Montreal Conference later that year the EC representative agreed to a 50 per cent reduction.

Developing countries successfully pressed for some concessions; they obtained a derogation that would allow them to increase their CFC use for the first decade of the agreement. They were also to receive technical assistance as well as information and advice on CFC recycling and conservation. The fear was that without some concessions Third World countries would set up their own CFC production plants.

But these concessions did not address all of the developing countries' major concerns regarding the evolving regime. Of particular concern was the fact that there were no provisions for a fund to help them make the switch to CFC substitutes. The fund was not established because the United States, Japan and the EC countries opposed it. Because there was no provision for financial assistance, three major developing countries—China, India and Brazil—refused to sign the protocol. These were important holdouts, because they had the potential to become major producers and consumers of CFCs. They refused to participate in the regime until the industrialised countries agreed to provide financial and technological assistance. Although, UNEP supported them in their calls for the establishment of the fund.

Scientific revelations about ozone depletion continued to unfold, and they spurred the regime strengthening process. The

Montreal Protocol came into force on January 1, 1989, and the negotiations on its amendment began promptly. At a meeting to amend the protocol in Helsinki in May 1989, the veto coalition began to shift its position dramatically. The EC members were among 80 nations voting for a complete CFC phase out by the year 2000.

At the London meeting, in June 1990, a new coalition of 13 industrialised states pushed for a 1997 deadline for the final elimination of CFCs. However, the four leading CFC-producing nations—the United States, the United Kingdom, France and Italy—joined by the Soviet Union, offered an alternate phase out date of the year 2000. The push by some industrialised countries for a 1997 deadline caused the United States to change from a lead to a veto role. This deadline has been extended by year 2010, but there is little hope of its complete implementation.

Developing countries were usually at a disadvantage with respect to the size and make-up of their delegation. A developing country was likely to be represented by a single generalist, whereas the industrialised countries were more likely to have a team of specialists. Countries that had a strong economic interest in the process were usually represented by large delegations. But, most delegations fell between these two extremes. They had, on average, two to four members, usually representing scientific, industrial, economic, legal and political disciplines. The delegations that had specialists had an advantage since these people usually had special knowledge of, or experience regarding, the ozone issue.

The potential for further regime transformation still exists. With the developing countries now on board, the regime can be further strengthened by speeding up the time table for phase out. There is thus need for greater cooperation between developed and developing nations in field of ozone layer protection.

Greenhouse Effect and Global Warming

The phenomenon commonly known as 'greenhouse effect' occurs due to emission of certain gaseous pollutants in the air, which, after the heating of the atmosphere, causes the average global temperature to rise. The effects produced by certain gases, such as carbon dioxide or water vapour, cause a warming of the earth's atmosphere by absorption of infrared radiation. The atmosphere protects us by serving as a light scattering and heat mediating blanket. A large portion of heat emitted from the earth is re-absorbed by the atmosphere and is, in effect, conserved, with the result that the surface of the earth is warmer than it otherwise would be. In simple words, this warming is called the greenhouse effect.

The atmospheric concentration of these gases has relatively increased since the advent of the industrial era. There has emerged a growing consensus in the scientific community that the anthropogenic greenhouse gas emission of carbon dioxide, nitrous oxide, CFCs, methane and low level ozone threatens to disrupt human societies and natural ecosystems.

The developed world accounted for nearly 80 per cent of the total industrial carbon dioxide emissions and the rest of 20 per cent originated in the developing world. On the other hand, 98 per cent CFCs are produced in developed and only 2 per cent in developing world. With the current trend of emissions, a WHO study speculates that the combined effects of the six most important greenhouse gases could by the early years of 2030s be equivalent to doubling of pre-industrial concentrations of carbon dioxide. The scientists are of the opinion that the earth will be hotter by 2° to 4°C by the middle of the 21st century.

The atmospheric blanket keeps the earth habitable and protects living forms from fatal radiations of the sun. The atmosphere intercepts solar radiation, reflecting a third of it, while the rest is absorbed by the atmosphere, ocean, ice, land and biosphere. The energy absorbed is balanced by the outgoing radiation from the atmosphere and the earth. This vital balance will be altered with infringement of greenhouse effect resulting in an overall rise in the average global temperature.

Amongst the most dramatic consequences of the greenhouse climatic disruption is an unprecedented rise in global sea levels to the extent of 8 to 25.5 inches by the middle of the 21st century if gas emissions continue at the present rate. Melting of polar ice and glaciers, as well as thermal expansion of seas, would cause worldwide coastal flooding and rise in sea level will result in submergence of many coastal areas. It will not only affect the

Figure 15.1

Greenhouse Effect

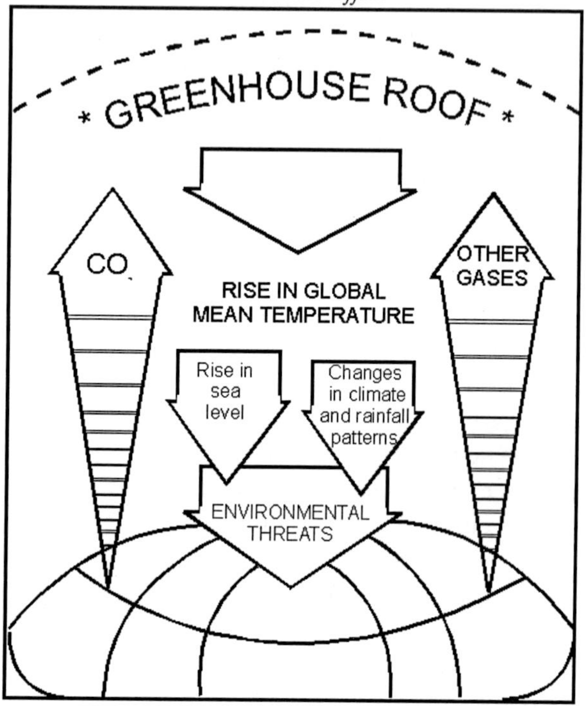

ecological balance but will also be a threat to the people and economy of the countries like Bangladesh, Netherlands, as well as to many coastal and island countries. In fact, the impact of world-wide climatic disruption could have hazardous for the survival of the planet both biologically and economically. The Montreal Protocol was a step in this direction but an overall solution requires combined efforts on the part of all nations and international bodies. Apart from this, scientific and technological research in this field can also reduce the risk to a great extent, provided it reaches all nations, specially to the developing nations.

Per Capita Greenhouse Index

National rankings reflect the contributions of individual countries to the atmosphere's growing store of greenhouse gases. But it is also of interest to compare countries by per capita contribution to these gases, since the citizens of our planet vary widely in their relative responsibility for these emissions.

Per capita figures can also be looked at as an indication of how much greenhouse gas emissions and global warming potential might grow if developing countries, in the process of development, significantly increase their per capita emissions. If just China and India, for example, raised their per capita emissions to the current world average, i.e., 1.1 metric tonnes of carbon dioxide heating equivalent per annum, total worldwide additions to the atmosphere would increase 28 per cent; and if to the level of France only, it would increase to 68 per cent.

Figure 15.2A compares the per capita emissions for the 10 countries of the Organization for Economic Cooperation and Development (OECD) with the highest National Greenhouse Index Rank. Similarly, Figures 15.2B and 15.2C compare per capita emissions for the 10 planned economies and 10 developing countries, respectively, with the highest national emissions. The large ranges among per capita emissions in each economic group suggest that national policies can make a significant difference in a country's contributions to atmospheric warming potential, although, geography also plays a role. Both the United States and Canada, for example, emit more than twice the per capita greenhouse gases that Japan does—a difference attributable to both geography and policy. Mexico emits about half the per capita emissions of Colombia and about one-fourth those of Brazil. At the same time, differences between economic groups are a reminder that any agreement to limit greenhouse gas emissions will have to take into account questions of individual equity and the legitimate aspirations of nations to develop.

The increase in atmospheric concentration of greenhouse gases is responsible for climatic change, especially in rise of temperature, or in other words, for global warming. Changes in greenhouse gas concentrations are closely correlated with changes in the earth's surface temperature. It has been established that warming of global climate and rise in CO_2 and CH_4 levels coincide. Some evidence

Figure 15.2
Per Capita Net Greenhouse Gas Emissions

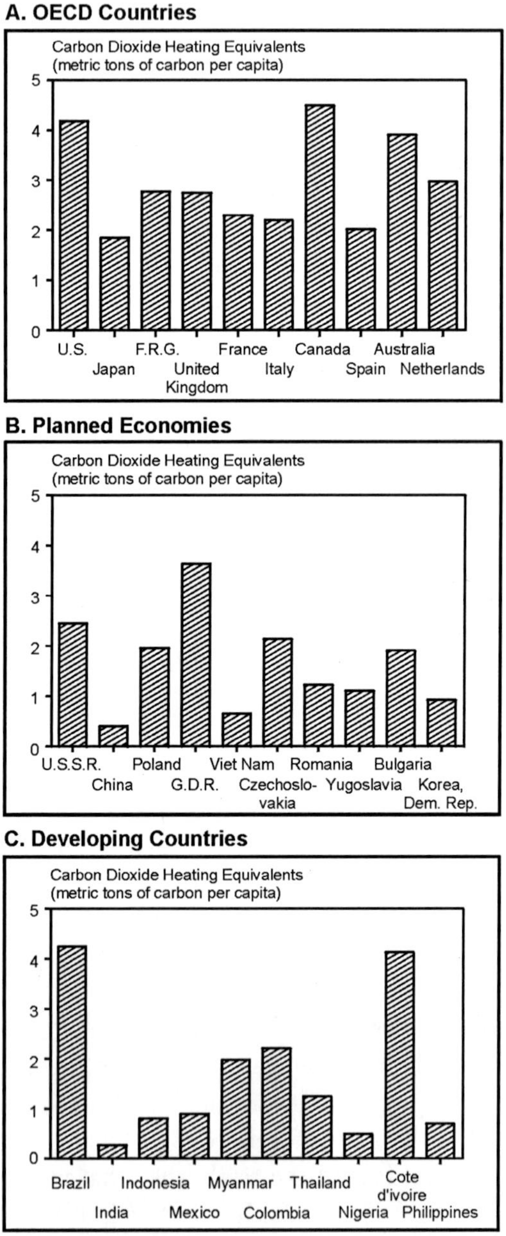

predicted that warming has begun to manifest itself already, but this evidence is controversial. Temperature records taken around the world since 1860 suggest that a global warming of 0.5°–0.7°C has occurred since this year, with nearly half of this rise occurring since 1965. Table 15.1 shows the contribution of greenhouse gases and human activity to global warming:

Table 15.1

Contributions of Greenhouse Gases and Human Activity to Global Warming

(per cent of global warming)

Sector	*Carbon Dioxide*	*Methane*	*Ozone*	*Nitrous Oxide*	*CFCs and others*	*Per cent Warming by Sector*
Energy						
Direct	35	3	×	4	×	49
Indirect	×	1	6	×	×	
Deforestation	10	4	×	×	×	14
Agriculture	3	8	×	2	×	13
Industry	2	×	2	×	20	24
Warming by Gas (%)	50	16	8	6	20	100

Source: V. Ramanathan, R.J. Cicerone, and H.B. Singh, "Trace Gas Trends and their Potential Role in Climate Change", *Journal of Geophysical Research*, Vol. 90, No. D3 (1985), pp. 5547–66.

× = not available.

Other Effects of Greenhouse Gases

Deep Ocean Currents

The disruption of deep ocean currents by global warming could have profound consequences on the global climate. These currents, which act as a great conveyor belt carrying roughly 20 times the volume of water flowing through all the world's rivers, depend on a delicate balance of salinity and temperature to drive them—a balance that may be shifted by warmer temperature in the polar regions and consequent changes in the amount of polar ice.

Evaporation from prevailing winds across the Atlantic brings rain to Europe and Asia, but raises the Atlantic's salinity. Chilled by Arctic cold in the North Atlantic, this salty water increases in density and sinks to the bottom of the ocean, where it travels south in a huge circuit that takes it around Africa and through the Indian Ocean to the Pacific Ocean. The Pacific's less saline waters make their way in a shallower current back along a roughly similar route that skirts several continents, ultimately balancing the oceans' salt budget and warming the European landmass in the process.

An infusion of less salty water, as might be brought on by the melting of Arctic ice, could cause the deep ocean currents to stall since surface waters could not gain the density needed to sink. Some suggest that this stalling of the ocean current systems has happened in the past, leading to dramatic effects on the climate. According to this theory, the rapid changes in climatic conditions that have signalled the end of recent ice ages may have been precipitated by shifts in dip currents.

The role of the oceans in taking up or supplying atmospheric CO_2 is another key uncertainty in the attempt to model greenhouse effects. Of the roughly 8.5 billion metric tonnes of CO_2 emitted yearly by humans from fossil fuel burning and deforestation, only about half remains in the atmosphere. The rest is taken up by the various links in the planet's carbon cycle, including land-based ecosystems and the oceans. The terrestrial biosphere, including all plants—alive and dead—contains about 2.5 times as much CO_2 as the atmosphere. But, the world's oceans contain about 50 times as much CO_2 as the atmosphere and, each year, over 200 billion metric tonnes cycle in and out of the oceans through gas exchange at the sea surface.

Cloud Behaviour

The behaviour of clouds in future greenhouse scenarios is also a major source of uncertainty in global climate models. Clouds play an important role in regulating the earth's energy budget. They routinely cover about half the earth and account for a good deal of the 30 per cent of indirect sunlight that the earth reflects into spaces. But clouds do not merely reflect solar radiation from the earth; they also absorb some of the heat radiated by the earth and reflect it back to the surface below. In other words, they also exert a greenhouse effect of their own. Currently, averaged over the globe, the reflective property of clouds dominates their greenhouse effect and, thus, they exert a substantial, net cooling effect on the earth.

Both the reflective effect and the greenhouse effect of clouds are large compared with the greenhouse effect exercised by CO_2 and other trace gases in the atmosphere. Recent data from a satellite-borne experiment show that, on a scale where the warming effect from CO_2 doubling is 1, the cloud warming effect is currently 7. Therefore, the result is a net cooling effect four time larger than the expected warming that would be caused by a doubling of CO_2.

Fertilisation Effect of Higher CO_2 Levels

Anticipating how this balance will change is complicated by the many competing influences that could pertain under greenhouse conditions. For instance, photosynthesis rates are known to be sensitive to atmospheric CO_2 level. The effect known as CO_2 fertilisation can cause an average 30 per cent increase in growth rates in many, but not all, species when CO_2 concentration is doubled. Recent data suggest that this effect is temperature-dependent and could be increased up to three-fold by a ·3°C rise in air surface temperatures.

Further, there is evidence that CO_2 enrichment helps some plants minimise water loss, presumably making them more drought-tolerant—an important consideration if the drier conditions predicted by climate models for many mid-continent areas come to pass. If all these favourable effects of greenhouse conditions were realised, a net increase in plant matter might increase the carbon uptake of the biosphere and act as a negative climate feedback, helping to alleviate further CO_2 build-up. Indeed, there is evidence that the annual carbon uptake in the Northern Hemisphere has increased during the growing season in recent years, perhaps indicating greater carbon storage.

This potential rise in methane production is particularly worrisome. Methane is 20–30 times more efficient than carbon dioxide at trapping heat, and its concentration in the atmosphere has been increasing more than 1 per cent per year. Within 50 years, it may be the most significant greenhouse gas. Large methane releases from a warmer tundra—or from other sources sensitive to warming, such as deep ocean deposits—could create a major climate feedback.

Biosphere Response

Predicting how living systems will respond to global warming and what climate feedbacks they will exercise may be just as difficult as solving the riddles of ocean dynamics or cloud behaviour and provide still another major source of uncertainty in climate modelling. Terrestrial ecosystems play an important role in the global carbon cycle, storing about 100 billion metric tonnes of carbon each year in plant materials through the process of photosynthesis and releasing about the same amount through the processes of respiration and decay. This annual flux is equivalent to almost 30 per cent of all the carbon held in the atmosphere, so even minor changes in the balance between photosynthesis, respiration and decay could affect the level of atmospheric CO_2 quickly.

Poleward Migration of Species

Perhaps, the most severe disruption of the terrestrial carbon cycle could come as result of the rapidity of the expected global warming. One consequence of this warming will be a movement towards the poles of the range of many plant species. But will plants be able to migrate poleward as fast as their ranges shift? Species migration is usually a rather slow affair, especially for many long-lived tree species. However, a rapid rise of 2°–5°C in global temperature in the next 100 years could well translate into a 500–1,000 kilometres range shift poleward—a distance that many species will not be able to travel. This may result in the gradual extinction of many forest communities and, perhaps, a concomitant decrease in carbon storage rates.

Other Uncertainties

In addition to the questions surrounding ocean, cloud and plant responses to global warming, there are other scientific uncertainties as well. Climate models do not describe sufficiently such important factors as the variation in solar output, the influence of volcanic and human-caused aerosols, the earth's reflectivity, and the influence of soil chemistry on the carbon cycle, as well as several other minor factors.

In spite of these many deficiencies, today's climate models provide a first estimation of global climate change with enough internal consistency and apparent relation to reality that they cannot be ignored, though their detailed predictions will require years of refinement.

Policy Options Available to Address Global Warming

The responses of human societies to global warming can take several approaches. First, prevention strategies can be employed to reduce the quantities of greenhouse gases being emitted. Mitigation mechanism strategies can attempt to compensate for emissions that do occur, for example, through reforestation policies that increase the uptake of CO_2 from the air by trees and other plants. Lastly, strategies can be employed that help communities and nations adapt to changes in climate and their consequences. In practice, all three types of approaches are likely to be important. Here, we discuss a range of prevention and mitigation policies.

It is worth noting that the problem of global warming is intimately linked to other serious environmental problems involving the atmosphere, in particular those of acid precipitation, urban smog, and depletion of the stratospheric ozone layer. The problems are linked chemically because, once released, many of the pollutants that cause trouble interact in complex and synergistic ways within the atmosphere or play a role in more than one problem. They are linked economically because these are often the same human activities that release the pollutants responsible for all three problems. And, they are linked in a policy sense because policies designed to attack one problem—by altering an economic activity to reduce emission of the pollutant responsible, for example—inevitably will affect other problems as well. A policy designed to produce automobile that use less fuel, for example, would reduce not only CO_2 emissions, but may also reduce the oxides of nitrogen (NO_2) emissions that contribute to acid precipitations, urban smog and global warming.

While work should proceed simultaneously on mitigation and adaptation strategies, prevention deserves the highest priority. Preventing the emissions of greenhouse gases that would be released into the atmosphere not only delays the onset of significant global warming, but also slows its advance and reduces its ultimate magnitude.

To reduce greenhouse gas emissions or even to slow their rate of growth on a global scale will require an extraordinary degree of political consensus. As the greenhouse gas indicators discussed earlier in this chapter make clear, no one country or even one region can prevent the build-up of greenhouse gases by itself, although leadership by individual countries will be important in achieving global consensus. To be broadly acceptable, ideally, prevention policies should confer other local benefits on their adopting countries in addition to the worldwide benefit of reducing the risk of a global warming.

Many policies have been proposed and are under study. It seems clear, however, that any successful prevention strategy will include five key elements:

- increasing the efficiency of energy production and use;
- switching from carbon-intensive fuels, such as coal to hydrogen-intensive fuels such as natural gas, where possible;
- encouraging the rapid development and use of solar and other carbon-free energy sources;
- elimination the production of most CFCs and developing the means to recapture those now in use; and
- reducing the rate of deforestation.

El Nino

The climatological event known as *El Nino* is a periodic rise in temperature of the eastern Pacific, around Peru and Galapagos Islands. There was a strong El Nino in 1982 and 1983 and a lesser version of the phenomenon was observed in late 1990s. In 1997 we have seen its impact on a worldwide scale. It was the most powerful

El Nino on record, resulting in heavy rains in the eastern Pacific and unusual dryness in Southern Africa and Central and South America.

In South-East Asia, these was a severe drought causing serious food shortage in New Guinea and contributing to forest fires in Indonesia, which enveloped the whole region in a choking smog. A subsequent warming of the Indian Ocean has lead to devastating floods in Somalia and southern Ethiopia. But far from being confined to the tropics, El Nino also had pronounced effects in Europe.

Scientists have been trying to come up with conclusive answers on the extent of damage El Nino would cause on the entire world. It has occurred twice in the last 15 years. Latin America, Cambodia and China and now the entire world is witnessing serious weather anomalies as an impact of El Nino, literally meaning 'little Christ' because generally it has occurred during Christmas.

Acid Rain

Acid rain is a matter of great global concern and has become one of the major environmental problems. The term 'acid rain', used first by Robert Angus Smith, the Chief Alkali Inspector of UK in 1872, described the acidic nature of rain falling around Manchester. Acid rain is commonly defined as "a condition in which natural precipitation becomes acidic after reacting chemically with pollutants in the air".

Among air pollutants sulfur and nitrogen are significant compounds. In particular, sulfur dioxide (SO_2) and oxides of nitrogen (NO and NO_2) are released from a variety of sources. Sulphur dioxide reacts in air to form sulphur trioxide. In turn, sulphur trioxide reacts rapidly with atmospheric moisture to produce sulfuric acid.

Similarly, the oxides of nitrogen react in air to produce nitric acid. Both these acids are very strong acids. These materials, when present in the atmosphere, dissolve in water droplets and fall to earth as acid rain.

Under normal conditions, rain water is slightly acidic because carbon dioxide dissolves in it and reacts to form a weak acid, which is not harmful. But man's contribution through fossil fuel, power plants, industries and automobiles has disturbed this natural acidic balance and converted natural rain into precipitated acid rain with several environmental implications.

The ill-effects of acid rain can be seen on vegetation, soil, marine resources, monuments as well as on man. Air pollutants and acids generated by industrial activities are now entering forests at an unprecedented scale. Many forests in Europe and North America now receive as much as 30 times more acidity than they would if rain and snow fell without pollutants. Its reaction includes change in colour of leaves, premature drops of leaves and flowers, tree crowns progressively thin, and, ultimately, trees die. Marine resources can be totally devastated with loss of aquatic life, fishes can die and entire lakes and streams can be destroyed. The affected fishes, if consumed by humans and birds, cause dangerous effects. In soil, the rate of decomposition of organic matter and formation of nitrogen fixing organisms is reduced by acids. Human health may also be affected by increasing respiratory and skin problems.

Acid rain is also an increasing problem in industrial areas of America, Europe and Asia, where large amounts of fossil fuels are burned. The areas affected by acid rain and potentially at risk have been depicted in Figure 15.4. The ill-effects of acid rains have no national limits as nature knows no geographical boundaries. Only developed technology can save the world from its harmful effects.

Desertification

The UN Conference on Desertification (UNCOD) has defined desertification as "the diminution or destruction of the biological potential of the land, which can ultimately lead to desert-like conditions". There is no environmental problem in the world that affects poor people as extensively as land degradation or desertification. According to the UN Environmental Programme (UNEP), about 900 million people are threatened by desertification, which affects more than 6.1 billion hectares, which about 35 per cent of the earth's land area. Each of the 16 years that have passed since the UN

Figure 15.3

Acid Rain from a Thermal Plant

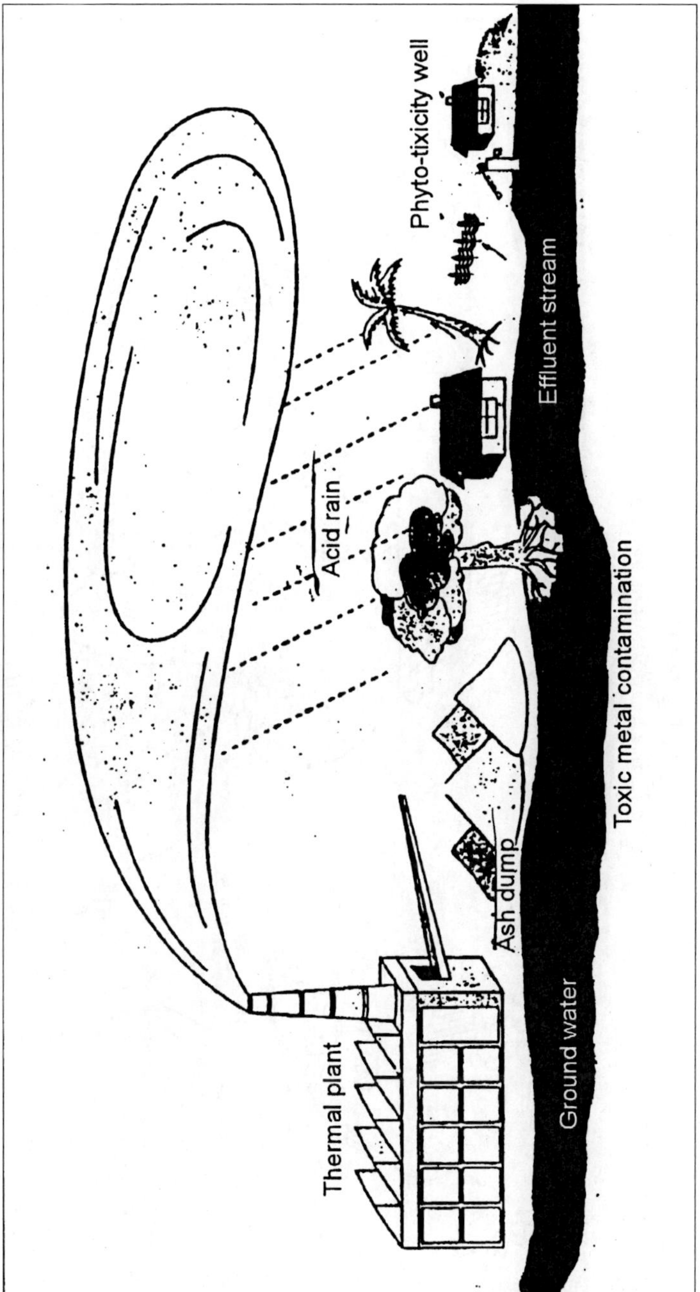

Figure 15.4

Acid Rain Distribution in the World

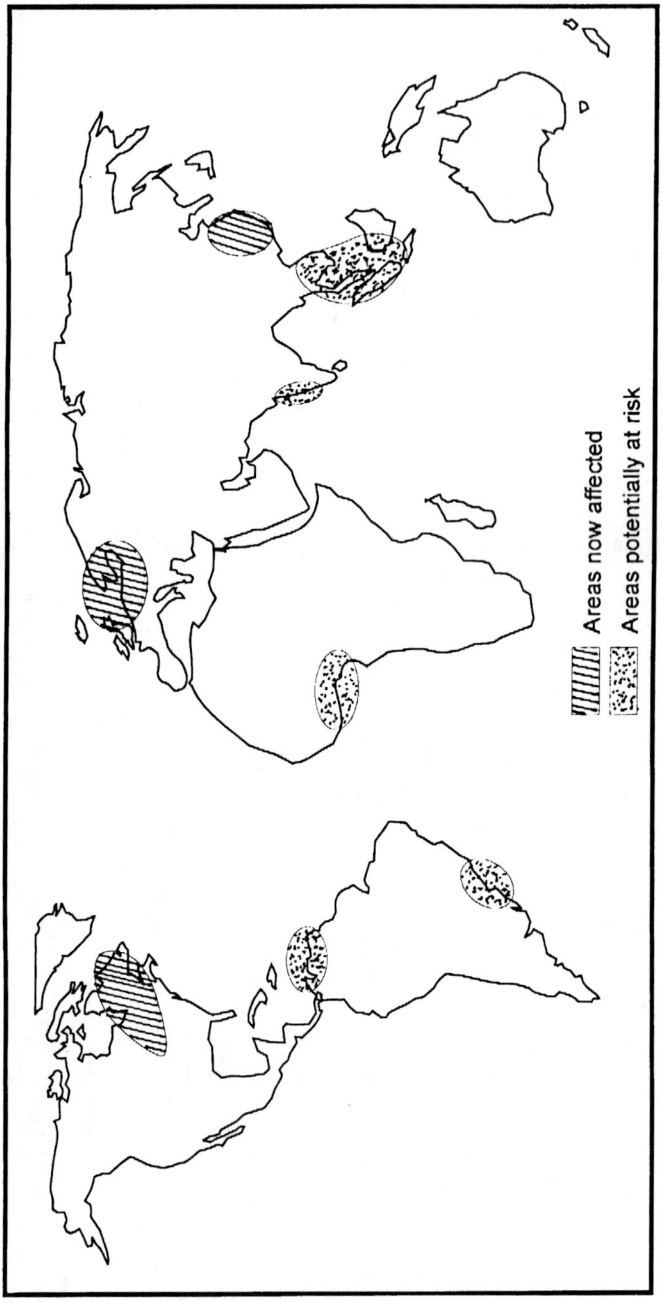

Conference on Desertification (UNCOD) in 1977 has seen approximately 6 million hectares of previously productive land rendered infertile. In terms of income, on an average, this is an annual loss of about 45 billion dollars.

Figure 15.5
Threats from Desertification

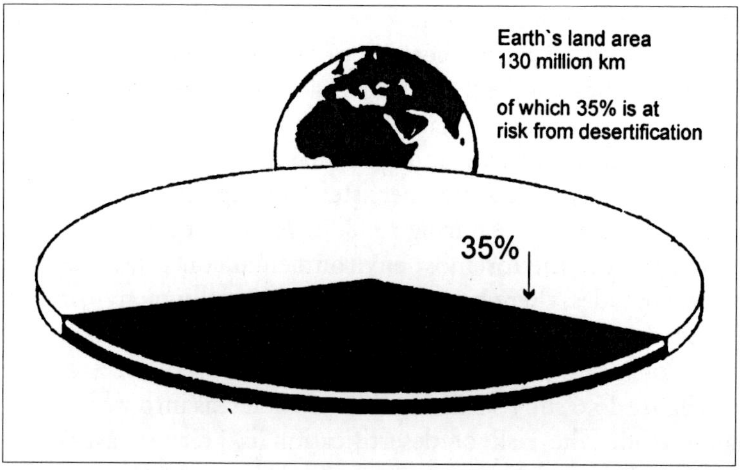

Desertification is the end product of a vicious mixture of economic, political and technological forces, which leaves people economically socially and politically marginalised. As the land gets further degraded, sheer survival takes up more and more human efforts, labour and time, till finally the men are forced to migrate and swell the ranks of the urban unemployed, while the women are left behind to struggle on.

Desertification is a slow process in which land productivity and resilience decline steadily. UNEP's latest study lists overgrazing, deforestation and unsustainable agricultural practices as the major causes of desertification. But, these practices may be motivated by a whole chain of economic practices, which result in poor prices for agricultural and livestock products, or political compulsions like debt, which force a country to promote adverse land use practices in order to earn foreign exchange.

What makes dryland soils specially vulnerable to degradation is the slowness of their recovery from disturbance. Because water is often not available or is available only in limited quantity, new soil is formed slowly, and salts, once accumulated, tend to remain *in situ*. Build up of organic matter, which is usually consumed very quickly under warm conditions, takes a long time. Moisture deficiency discourages recolonisation by plants that have been removed or damaged.

Asia, with 1.3 billion hectares of degraded drylands, suffers the most desertification. But, in terms of severity, Africa and North America are worse off, because nearly three-fourth of drylands are affected. In fact, desertification is a major environmental problem and has worsened by the persistent droughts reported in the Northern Hemisphere during 1988. In Africa, drought and desertification rank as the foremost environmental concerns. During the last two decades, there have been at least 10 million environmental refugees from the drylands. Desertification is a matter of global dimensions.

Figure 15.6 shows existing deserts and areas into which deserts may spread. The risk of desertification is great in areas where rainfall is low and evaporation is high. However, the politics on desertification have turned full circle after the Intergovernmental Negotiating Committee for Convention to Combat Desertification (INC-D) held in May 1993 in Nairobi. To check desertification it is necessary that nations make combined efforts and if necessary consumers should pay ecological costs of the products exported from developing countries. Only international efforts can only check or minimise the process of desertification.

Deforestation

Forests are not merely a natural resource to supplement multiple human requirements but are a very vital component of the environmental system. The forest plays a pivotal role in balancing the ecosystem. Growing industrialisation, urbanisation and ruthless exploitation of forests have created chaotic conditions and severe phyto-geographical and environmental imbalances. The use of forests has been linked with mankind from times immemorial

Figure 15.6
Earth's Spreading Deserts

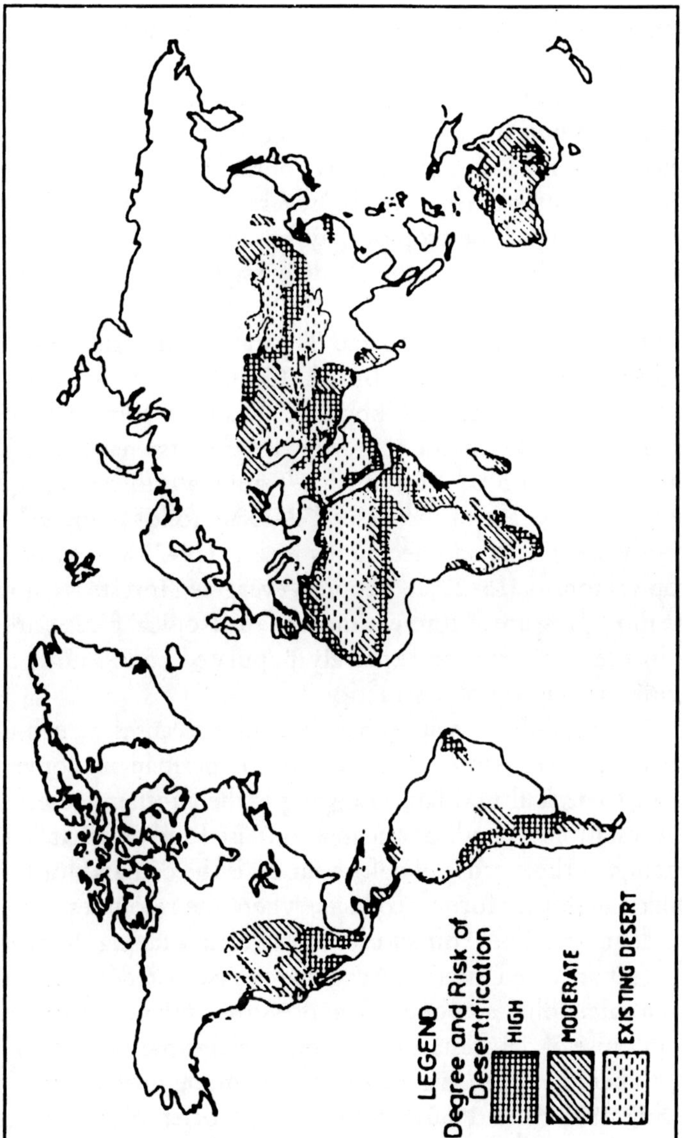

basically for food, shelter and clothing, but population exodus and stresses on forest resources created a condition where natural growth of forests and their maintenance is incapable of coping with the demand.

Tropical forests are the richest forest resources of the world, their distribution has been depicted in Figure 15.7. But tropical deforestation has become an environmental issue over the past few years. Recent studies covering several key countries suggest that deforestation in the tropics may be much worse than was previously thought. The annual deforestation in the tropics was estimated to 11.4 million ha in 1980 but now it has been estimated to be around 20.4 million ha. The rate of deforestation is much higher in countries like Brazil, Cameroon, Costa Rica, India, Indonesia, Myanmar, Philippines, Thailand and Vietnam. Since pre-agricultural times, the world's forests have declined by one-fifth, from 5 to 4 billion ha. Temperate forests have lost the highest percentage of their area (32–35%), followed by sub-tropical woody savannahs and deciduous forests (24–25%) and old growth tropical forests (15–20%). Tropical evergreen forests are now under the most pressure. Although most of the tropical forests are located in inaccessible areas and sparsely populated zones, now, these are facing the tragedy of destruction.

Tropical deforestation has three direct causes, which often act together in the same area. The first is permanent conversion of forest to agricultural land. Logging is the another cause, although deforestation depends on the methods used and on what follows the logging. A third cause of deforestation is the demand for fuelwood, fodder and other forest products, where the resources cannot meet the demands. These direct causes as well as grazing, forest fire and drought are often interrelated and are exacerbated by government's economic policies, population pressure and poverty. Thus, the main cause of deforestation is commercial exploitation of forests.

There has been growing concern among professional foresters during 1970s and 1980s about the rate of tropical deforestation. With the introduction of the Tropical Forestry Action Plan (TFAP) in 1985, FAO, UNDP, World Bank and other government and non-government organisations have expressed the opinion that

Figure 15.7
Global Distribution of Tropical Forests

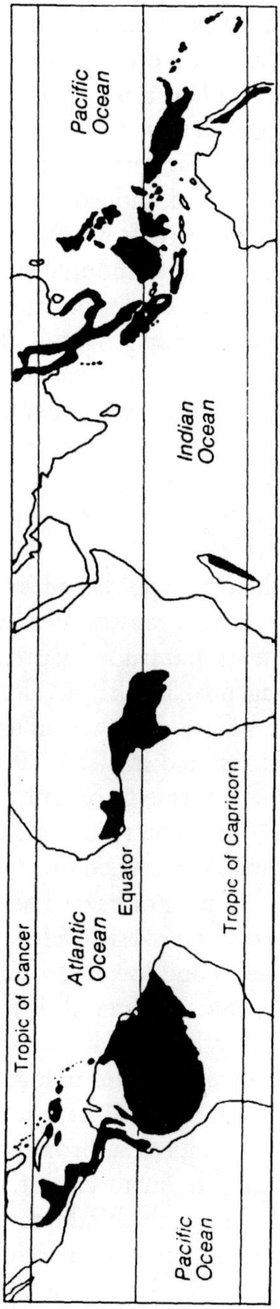

efforts to improve forest management had failed, because many causes of deforestation lie outside the domain of national forest departments. The problem must therefore be addressed at higher political and economic levels. TFAP was designed to deal not only with deforestation but also with forest degradation, with emphasis on agro-forestry, fuelwood, industrial forestry, and other methods by which forestry can provide goods and services to rural people and contribute to national economy. More than 60 countries have decided to prepare national forestry action plans to guide management of their forests. As the role of tropical forests in tempering global warming and harbouring a diverse range of species has become clear, environmental groups have pressured international developmental agencies to halt projects that cause deforestation.

Hazardous Waste Trade

In the late 1980s, the media published a spate of news stories about the dumping of hazardous wastes in the developing world. In the wake of these incidents, hazardous waste trade moved on the top of the international agenda. UNEP accelerated talks addressing the regulation of this trade, and the world's nations debated the issue. These negotiations resulted in the 1989 Basle Convention, which established a 'prior informed consent' system. The convention, adopted in March 1989, entered into force in May 1992. It was strengthened in March 1994, when the contracting parties agreed to a complete ban on shipping of hazardous waste.

The term 'hazardous waste' refers to material that poses a threat to human health and the environment, especially if it is not properly managed. Such material includes radioactive wastes, polychorinated biphenyl, pesticides, toxic incinerator ash, chemical sludge and organic solvents. An estimated 90 per cent of the world's hazardous wastes is generated by industrialised market economies. The United States produces approximately 80 per cent of the world's total output. The main exporters of hazardous wastes in western Europe have been the Netherlands, Belgium, France, Italy and Germany. All these countries handle the major proportion of the wastes they generate. The United States handles the major proportion of its wastes domestically, exporting only 1 per cent. About one-fifth of the total annual global trade in wastes goes from industrialised countries to developing countries, including India. Most of the Third World countries lack the technology or administrative capacity to treat or dispose of the wastes safely. The only attraction of developing countries has been the low cost.

Although the bulk of the hazardous waste trade has occurred between developed countries, the issue gained on the international agenda because of the increasing numbers of illegal hazardous waste shipments to developing countries. The Basle Convention is the landmark on this issue. Its main provisions are:

- Hazardous waste generation is to be reduced to a minimum.
- It provides for prior informed consent.
- For any trans-boundary movement, it requires environmentally-sound disposal of hazardous wastes.
- Any state has the sovereign right to ban imports.
- Convention obligations apply to both recycling and disposal operations, since it also has the potential for environmental damage.
- Developing countries are entitled to receive assistance from developed countries that are parties to the convention, etc.

Biodiversity

Nature and Importance

Biodiversity or biological diversity refers to the variety of life on earth. The term encompasses three categories: genetic diversity, species diversity and ecosystem diversity. 'Genetic diversity' describes the variation of genes within a species, 'species diversity' describes the variety of species within region, and 'ecosystem diversity' refers to the number and distribution of ecosystems. The increasing interest in biodiversity is the result of concerns regarding species extinction, depletion of genetic diversity, and disruptions to the atmosphere, water supplies, fisheries and forests.

The environment is a complex of ecosystems made up of a variety of species, themselves represented by a number of strains, each with is own genetic identity. This diversity is a condition for the long-term sustainability of the environment and the maintenance of its integrity is, therefore, recognised as being indispensable to sustain human life. The taxonomic knowledge of species is incomplete and, therefore, it is essential to guard against the extinction of the largest possible number of species and strains, some of which, in the future, may provide us with food and medicine or can be used in the biological control of pests and pathogens. Biological diversity encompasses all species of plants, animals and micro-organisms and the ecosystem and ecological processes of which they are part. Genetic diversity is the sum total of genetic information, contained in the genes of individual plants, animals and micro-organisms that inhabit the earth. Species diversity refers to the variety of the living organisms on earth. Ecosystem diversity refers to the variety of habitats biotic communities and ecological processes in the biosphere as well as the enormous diversity within ecosystems in terms of habitat difference and the variety of ecological processes.

Biodiversity provides the basis for life on earth. The fundamental social, ethical, cultural and economic values of these resources have been recognised in religion, art and literature from the earliest days of recorded history. Wild species and the genetic variations within them make substantial contribution to the development of agriculture, medicine and industry. Perhaps, even more important, many species have been fundamental to stabilisation of climate, protection of watersheds, protection of soil, and the protection of nurseries and breeding grounds. It is difficult to determine the total economic value of the full range of goods and services which biodiversity provides.

Biological resources have made substantial contribution to human welfare, specially in the rural areas of developing countries. For example, fuelwood and dung provide over 90 per cent of the energy needs in many areas in Nepal, and in many other countries. Human beings have used about 7,000 kinds of plants for food, but only 20 species supply 90 per cent of the world's food, and just three—wheat, maize and rice—supply more than half. Although the major food crops in use today were domesticated thousands of years ago, the potential for other species to play prominent role in world agriculture is great. Numerous locally important, though yet much less known species of wild and domesticated plants could be cultivated and used much more widely.

Genetic diversity of a crop is the total genetic variation of the species available for further crop improvement. With and ever-increasing growth in human population, more food has to be grown to satisfy the needs. The key to better yielding varieties is provided by plant breeding to increase their food value, resistance to fungal, bacterial and viral diseases and insects, nematodes and other pests. This inbuilt genetic resistance and adaptation may eliminate the need for environmentally undesirable pesticides.

More than four lakh varieties of germplasms of important crops have been collected in international gene banks. Some countries, including India, have specific crop gene banks. There are numerous facilities in the world that conserve collection of vegetatively propagated crops like fruit trees, tuberous crops, palms, nuts coffee, cocoa, grasses and shrubs. World collection also include varieties of wheat, rice, maize, barley, potatoes, sweet potatoes, sorghum, millets, lentils, legumes, etc.

Estimation of World Species

How many species exist in the world is a big question before the scientists and they have given their estimates, but still require a long way to go to unfold the entire biodiversity pattern of the world. So far, about $1.6x10^6$ species have been described and of those fewer than 10^5 are familiar, interesting, or pretty enough to have been studied in detail. More are being discovered all the time. In 1988, two primates and one deer were discovered, another primate (a tamarin) was found in 1990, and since 1908, 11 species of cetaceans (whales and porpoises) have been added to the global inventory, amounting to 13 per cent of all known members of that order. Between 1978 and 1987, the newly identified animal species were 5 birds, 26 mammals, 231 fish, and 7,000 insects. Clearly, the species that have been found represent only a proportion of the total number, and estimating that number must necessarily involve an extrapolation from what is known to what is unknown. Many scientists have attempted the task, producing estimates of up to

100×10^6 species, but a conservative and probably more realistic estimate puts the total at $5\text{–}10 \times 10^6$. Robin Pellew, director of the UK branch of the World Wide Fund for Nature, has taken a median estimate of around 8×10^6 broken down as in Table 15.2.

Table 15.2

Number of Species Described and the Likely Total Number

Group	*No. Described (thousands)*	*Estimate of Total No. (thousands)*
Viruses	5	500
Bacteria	5	400
Fungi	7	1,000
Protozoa	40	200
Algae	40	200
Nematodes	15	500
Molluscs	70	150
Crustaceans	40	100
Arachnids	75	600
Insects	950	4,000
Vertebrates	45	50
Higher plants	250	300
Total	1,605	8,000

The exact number is likely to be higher, because the list omits a number of invertebrate groups, such as the annelids (worms), sponges (porifera) and cnidarians (hydras and jellyfish). It also omits such plant groups as the ferns, mosses and bryophytes. Further uncertainties arise from taxonomic disagreement between 'lumpers' and 'splitters' that can alter significantly the number of species recognised in certain groups.

Only 13 per cent species are supposed to be known of the world's existing microbial resources covering algae, bacteria, fungi lichens viruses and protozoa. Diversity of micro-organisms is preserved through culture collections, to be used beneficially in mining for metals, getting rid of methane from coal mines, cleaning up oil spill, creating perfumes, monitoring air pollution, controlling insect pests, destroying patricides in the soil, etc.

Many plants and invertebrate species, particularly associated with tropical forests, are still not identified and remain unknown. Most developing countries are concerned about how to safeguard their interest so that these resources are not lost to the technologically and financially more powerful developed countries. And to do this, the genetically rich developing countries, including India, must develop technological capabilities to safeguard the migration of vast germplasm resources through international agreements or financial attractions.

Like other natural resources, this distribution of living species in the world is not uniform. Species richness increases from the poles to the equator. Freshwater insects, for example, are three to six times more abundant in tropical areas than in temperate zones. Tropical regions have also the highest richness of mammal species, per unit area, and vascular plant species diversity is much richer at lower latitudes.

Extinction of Species or Loss of Biodiversity

The extinction of several species or loss of biodiversity is a major environmental problem. Species are disappearing rapidly and there is growing consensus within the international community that a system should be put in place to slow or halt the process of extinction. All the species are the integral part of ecosystem and extinction of some species diminishes the well-being of the remaining species, including human beings. The concern about biological diversity is the shrinking genetic pool. It is estimated that tropical forests contain at least 50 per cent and perhaps 90 per cent of the world's species. According to an estimate, 20 to 75 species are becoming extinct each day because of deforestation in the tropics.

Species do not persist indefinitely. Indeed, it was the realisation that fossils are the remains or traces of organisms no longer to be found alive that led to the development of our present understanding of evolutionary processes. Estimating the number of species existing at present as a proportion of all the species that have ever lived is difficult, but the plants and animals alive today probably constitute around 2–4 per cent of all the species that have lived in the last 600 million years.

Extinctions have not been spread evenly. There is a general background rate at which species go extinct, but there have also been at least five mass extinctions in which 65–85 per cent (and at the end of the Permian period, around 225 million years ago, 95 per cent or more) of all species disappeared. Background extinctions have little or no effect on biodiversity, because their rate is matched or exceeded by that of speciation, in which one species divides into two and thus increases the number of species. Nor do mass extinctions produce more than a temporary reduction in the number of species, because historically they have been followed by the rapid adaptive radiation of surviving species into the vacated niches. The extinction of the dinosaurs, for example, was followed by rapid speciation among mammals. Today, the number of land-dwelling species alive is about twice the average of the last 450 million years.

Throughout the geological history of the earth, species of plants and animals had been subjected to various evolutionary processes. Some species became extinct during the different geological periods. The last major extinction of some species occurred at the close of the Cretaceous period, some 65 million years ago, when birds and mammals were particularly affected. The total disappearance of all dinosaurs occurred during that great extinction. It is generally agreed that they were triggered by natural environment phenomena during geological evolution of the earth. In recent history, biological resources have been lost at an accelerated rate, mainly due to anthropogenic causes. Tropical deforestation between 1990 and 2020 may eliminate between 5 and 15 per cent of the world's species. The World Conservation Monitoring Centre has recognised that some 22,000 species of plant and animals are actually threatened by extinction.

Amongst the causes identified for the loss of biological diversity are: (i) habitat loss, fragmentation and modification: as forests are cleared, wetland drained, valleys flooded and roads built, so habitats are transformed and effectively lost for certain species; (ii) overexploitation of resourcees; for example, commercial harvesting has been a threat to many marine species, and also extinction of some large terrestrial animals, like the African elephant; (iii) pesticides and pollution have affected several species of birds and other organisms; and (iv) the impact of introduced exotic species as the threaten natural flora and fauna by predation, competition or altering natural habitat.

The introduction of new high yielding wheat and rice varieties, since the mid-1960s, has caused a loss of gene pool in many centres of crop diversity like Turkey, Iraq, Iran, Afghanistan, Pakistan and India.

The intense pressures on biological diversity are a direct reflection of the increasing human population. These pressures are expected to increase until population stabilises, as projected by the United Nations, by during the 2050–2070 period at about 10 billion. Such stabilisation would be achieved only if present efforts to curtail growth of human population are pursued vigorously.

Evolution of the Biodiversity Regime

The Stockholm Conference was not the first occasion at which issues related to biodiversity were placed on the international agenda. A variety of legal instruments already existed. Some of these were regional, such as the Convention on Nature Protection Wildlife Preservation in the Western Hemisphere (Washington, D.C., 1940) and the African Convention on the Conservation of Nature and Natural Resources (Algiers, 1968). Other measures dealt with specific species, such as birds and marine and polar region species. Prior to Stockholm, the only major international legal instrument for conserving biodiversity was the 1971 Convention on Wetlands of International Importance Especially as Waterfowl Habitats. The 61 parties to this convention pledged to make an inventory of designated wetlands, to promote their conservation, and to establish nature reserves in the important wetlands.

But, no comprehensive international treaty addressed the problem of biological diversity. Although Stockholm did not produce such a document, it was the first time the issue was addressed in such a holistic manner. The issue of biodiversity was debated, and the conference documents included provisions aimed at protecting biodiversity

The area of genetic resources and biotechnology was already controversial in 1972, at the time of the Stockholm conference. Developing countries were concerned about their lack of access to the biotechnology that was developed from the use of their resources, and they tried to use linkage politics to gain access to this technology. They pressed their demand for the transfer of technology on non-commercial terms. This demand was rejected by the developed countries, and the final statement of the conference diluted the language that addressed this linkage between economics and the environment.

Both the declaration of the United Nations Conference on the Human Environment and the Action Plan for the Human Environment included language that addressed biodiversity. Principles 2, 4, and 6 of the declaration included provisions for safeguarding flora and fauna and preserving ecosystems. The Action Plan generally addressed the importance of protecting forests, which among other thing, were important habitats for wildlife; it recommended that wildlife be monitored to assess the impact of pollutants and that the economic value of wildlife be assessed. Governments were asked to take steps to protect ecosystems that have international significance. Recommendations 39 through 45 deal generally with the issue of preserving genetic resources.

In the period between the Stockholm and Rio conferences, the body of legislation concerning biodiversity continued to grow. Relevant major international legislations included the convention concerning the Protection of the World Culture and Natural Heritage (1972), the Convention on International Trade in Endangered Species of Wild Fauna and Flora (1973), and the Convention on Conservation of Migratory Species of Wild Animals (1979).

In the aftermath of Stockholm, developing countries continued their push for greater access to the biotechnology that resulted from their biological wealth. A non-binding International Undertaking on Plant Genetic Resources (Resolution 8/33) was passed by the FAO in 1983. It stated that access to genetic resources should not be restricted, and it declared that all seed resources were the common heritage of humanity; therefore, both primitive stocks and those developed by proprietary means should be free to all. Some developed world producers were opposed to this, and many industrialised countries were not parties to it because it was incompatible with their patent rights. Third World members, however, accepted its provisions.

A 1991 Costa Rican deal is considered one possible model for future bilateral arrangements. The parties to the deal were the National Biodiversity Institute, a non-profit Costa research centre, and Merck & Co., the world's largest pharmaceutical company. Merck agreed to pay the institute US$1.1 million, as well as royalties from any product developed, in return for plant, insect and microbe samples from all over Costa Rica. Costa Rica is estimated to have about 5 per cent of any royalties go directly into conservation. Although this agreement provides for compensation, some critics say it gives away Costa Rican resources for too little money. Nevertheless, many developing countries are studying this arrangement with interest.

The years between Stockholm and Rio saw incremental progress with regard to the items on the Third World's biodiversity agenda. Developing states wanted to make their own arrangements concerning resources access, and as the Merck deal illustrates, the slow process of norm change had begun. But, the progress of these states in the campaign for transfer of technology and financial assistance is more difficult to assess. Although developing countries continued to push for change in these areas, developed countries maintained their determined resistance. These conflicting perspectives were reflected in the preparatory negotiations for the Rio conference.

The declaration of UN Conference on the Human Environment and the Action Plan for the Human Environment addressed the importance of protecting forests, which among other things, are important habitats for wildlife. Other conventions of biodiversity include May 1992 at Nairobi; June 1992 at UNCED; while in December 1993 Convention on Biological Diversity entered into force. The highlights of the convention were:

(i) National identification and monitoring of biological diversity.

(ii) The development of national strategies and programmes for conserving biological diversity.
(iii) National *in situ* and *ex situ* conservation measures.
(iv) Environmental assessment procedures to take into account the effects of projects on biological diversity.
(v) National reports from parties on measures taken to implement the convention and the effectiveness of the measures.

The issue of biodiversity should be seen as a common problem; as such, action by any country has implications for any other country. The biodiversity of every country should be protected with mutual cooperation and exchange of biotechnology.

In fact, this is a common property resource concern, as decrease in biodiversity has consequences for all species. But, this common property resource has some distributional characteristics that makes it different from other environmental issues. The bulk of the world's biological diversity is found in tropical forests, which are located in the Third World. The loss of biological diversity has tremendous implications for agriculture. The range of food resources has declined over time. One estimate says that more than 3,000 plant species have been used for food throughout human history, however, there certainly has been a significant reduction in variety. Now, most of the world's food comes from 20 species. The issue of patents is at the core, nowadays, because several developing countries have a fear that the use of traditional plant species will be restricted by scientifically advanced countries. After Uruguay round of GATT seeks to standardise intellectual property rights, much debate has been done and still continues. The controversy over patent right of 'Neem' tree, which has been used by Indians for centuries, is one example.

To conclude, one can quote from a stanza in *Isho-Upanishad* which says: "The Universe along with its creatures belongs to the land. No creature is superior to any other. Human beings should not be above nature. Let no one species encroach over the rights and privileges of the other species."

Adverse Effects of Insecticides and Pesticides

The excessive use of chemical pesticides and insecticides has resulted in degradation of environment, which has become a threat to human health. Throughout the ages, farmers have had difficulty in protecting their crops from small animals and disease organisms. At times, pest species have boomed and brought disease and famine. But, most of the time, natural balance has been maintained, and humans have lived together with insects in reasonable harmony. Nowadays, a great debate over the use and misuse of insecticides in agriculture is going on. On the one hand, scientists claim that more production can only be possible through control over pests and weeds, while on the other hand, they disagree and feel that chemical pesticides are harmful to the environment and actually increase pest problems.

The modern pesticide era started in about 1940 when a chemical called *DDT* was discovered to be a potent insecticide. DDT was cheaper and more effective than the previously known chemicals and led to dramatic early successes. But this miracle chemical had fallen from grace and on 1 January 1973, all inter-state sale and transport of DDT was banned in the USA, ban was also imposed on several other pesticides like aldrin, clieldrin, etc. Most of the insecticides are broad spectrum poisons and their use has led to disruption of ecosystems. Because these materials are not biodegradable, they persist for long periods in the environment. In some soils DDT and aldrin have been detected 15 years after a single spray application. Such persistence leads to a series of several ecological problems. All insecticides have had adverse impact on human health and since the mid-1940s, many thousands of people have been sickened or have died from severe pesticide poisoning every year.

Although the use of DDT in the US and other developed countries, was banned in 1973, it is still being manufactured and exported for use in other countries, specially to Third World countries. Thus, long-term global contamination continues. What is needed is the use of organic fertilisers and natural system of control through certain bacteria. Pests can be controlled without killing them directly. There is a need for integrated pest management and to evolve scientific methods to control pests and weeds without adverse effects on ecosystem. In addition, government restrictions may serve as a drive to develop other, more environmentally sound methods of pest control.

Pollution of the Oceans

For many years, people were relatively unconcerned about the pollution of the oceans because the great mass of the sea can dilute a huge volume of foreign matter to the point where it has little effect. In recent years, this attitude has changed because pollutants are often added in such relatively high concentrations that environmental disruption does occur before the pollutants have a chance to disperse. In addition, the quantity of pollutants dumped into ocean has grown so large that some scientists fear that global effects may be significant.

Oil spilling in oceans is a major problem. Between 1969 and 1983 there were more than 500 tanker accidents that involved oil spills. Altogether, more than one million tonnes of oil was released. During Gulf War a huge amount of oil spilled in nearby seas and caused great damage to the ocean life. Apart from accidents, seepage of oil is also common. Sometimes thousand tonnes of oil spills out of the tanker by large leakage as happened recently (October, 1997) near Singapore. Even tanker captains often clean the oily holds of their ships in ocean waters.

Offshore drilling operations also contribute their share of oil to the sea. The largest spill so far started on June 3, 1979, at an offshore oil well owned by Pemex, the Mexican National Oil Company. The total quantity of oil that finds its way into the sea each year is very large. There is a tendency to dispose of high poisonous chemical wastes including chemical warfare agents, pesticide residues and radioactive substances into oceans. The sealed drums are usually dumped into the sea. All these dumping activities have had adverse effect on the entire ecology of the oceans. During 1972 Stockholm Conference on Human Environment, a study entitled, 'The Health of Oceans' was released, which has attracted the attention of world leaders but till now nothing substantial has been done in this field. The blasting of atom bombs under ocean waters is the greatest threat to the ocean environment, which must be banned immediately by all nations.

Environmental Warfare

Using weather as a weapon and manipulating the environment for military purpose is common and has come to focus in the Gulf

War. The United States was the first to use weather modification as a weapon in Indo-China region from 1967-72. The intention was to increase normal monsoon rainfall and disrupt Vietnamese logistics marking the emergence of the use of environment as a weapon in warfare. Artificial triggering of earthquakes and tidal waves, modifying hurricanes or hailstorms to change their course, creating artificial fog to impede enemy traffic, etc.

The use of chemical weapons is not only harmful to people and other organisms but to ecosystem also. The emergence of cruise and ballistic missiles and other types of bombs are responsible for a lot of pollution in atmosphere. The Gulf crisis (1990) has brought it back to the forefront. Dumping millions of barrels of crude oil in the sea to hamper naval operations and foul up desalination plants supplying drinking water, marked the beginning of environmental warfare. Military experts are of the opinion that every state has its environmental or geophysical vulnerabilities that can be exploited by an enemy in the event of war. All this can be checked through awareness and collaboration at international level.

In brief, global environmental issues discussed in this chapter highlight that now it is high time to protect the biosphere from environmental catastrophes. These environmental issues have become a part of discussion not only at environmental conferences but also in every international political meet. Several documents and agreements have been signed between nations. It will not be enough to have the agreement, but their implementation is also necessary at all levels. The growing globalisation might make the task even more difficult. With this, the rich have greater freedom to exploit the poor. With its attendant economic asymmetry, the globalisation of the economy has clear consequences for environmental health and sustainable development. There is thus need to evolve a global strategy, which should be supported by action at all levels, especially at the community level.

16

Environmental Problems and Planning in India

A country's environmental problems are affected by the level of its economic development, the availability of natural resources and the lifestyle of its population. In India, rapid growth of population, poverty, urbanisation, industrialisation and several related factors are responsible for the rapid degradation of the environment. Environmental problems have become serious in many parts of the country, and can no longer be neglected. The main environmental problems in India relate to air and water pollution, particularly in the metropolitan and industrial zones, degradation of common property resources which affect the poor adversely due to a degeneration of their life support system, threat to biodiversity and inadequate system of solid waste disposal and sanitation with consequent adverse impact on health, infant mortality and birth rate. The degradation of natural resources imposes several ecological problems. In earlier chapters of this book, problems of water, air, noise and soil pollution as well as their impact of human activities on environment in India have been discussed in brief. But their overall analysis is necessary not only for their understanding but also for the environmental planning in India.

Air Pollution

Besides some natural process, the main sources of air pollution in India are automobiles, thermal power plants, industries, agriculture and quarrying. During the process of combustion of fossil fuels, oxides of carbon and particulates are generated. Apart from this, various industries also generate other pollutants such as SO_2, H_2S, ammonia, oxides of nitrogen, hydrocarbons, and other toxic substances, including ozone, lead and fluorides. The nature of urban and rural air pollution is quite different in India, therefore, their respective causes, effects and remedies are also different.

Urban Air Pollution

Urban air quality has generally been deteriorated throughout the Indian cities. The reasons are rising number of motor vehicles, especially poorly maintained, and growing industrialisation without any priority for pollution abatement. Indian cities are considered as some of the most polluted cities in the world and on the top of the list is the national capital, New Delhi. The single most factor responsible for this deterioration of air quality in cities is the exponential increase in petrol and diesel fuelled vehicles. The number of vehicles have been growing at a rapid pace of 12 per cent per annum and their total has crossed a 50 million mark. Of these 70 per cent are two and three-wheelers with inefficient and highly polluting two-stroke engines. The number of two-wheelers has been growing at a rate of about 20 per cent per annum and their number was around 36 million by the turn of the 20th century. It was also expected that passenger cars and diesel vehicles will also be double by that time. The percentage of vehicular population in six Indian metros has been depicted in Figure 16.1.

It also becomes clear from Table 16.1 that Delhi has the maximum vehicular pollution with total emission of 872 tonnes per day, followed by Mumbai having 549 tonnes. The concentration of pollution is much higher at the highways, road signals and traffic junctions.

Industries are also responsible for the air pollution in India. Most of the industries are located near cities, irrespective any environmental consideration. They have not even adopted normal pollution control measures, therefore, are the main cause of air pollution. For example, Mumbai has various industries, mainly chemical (523), textile (531), and mechanical workshops (3,348). Besides, there are 32 factories processing rubber or plastic products.

Figure 16.1

Vehicular Population and Percentage of Categories of Different Types of Vehicles in Major Metropolitan Cities

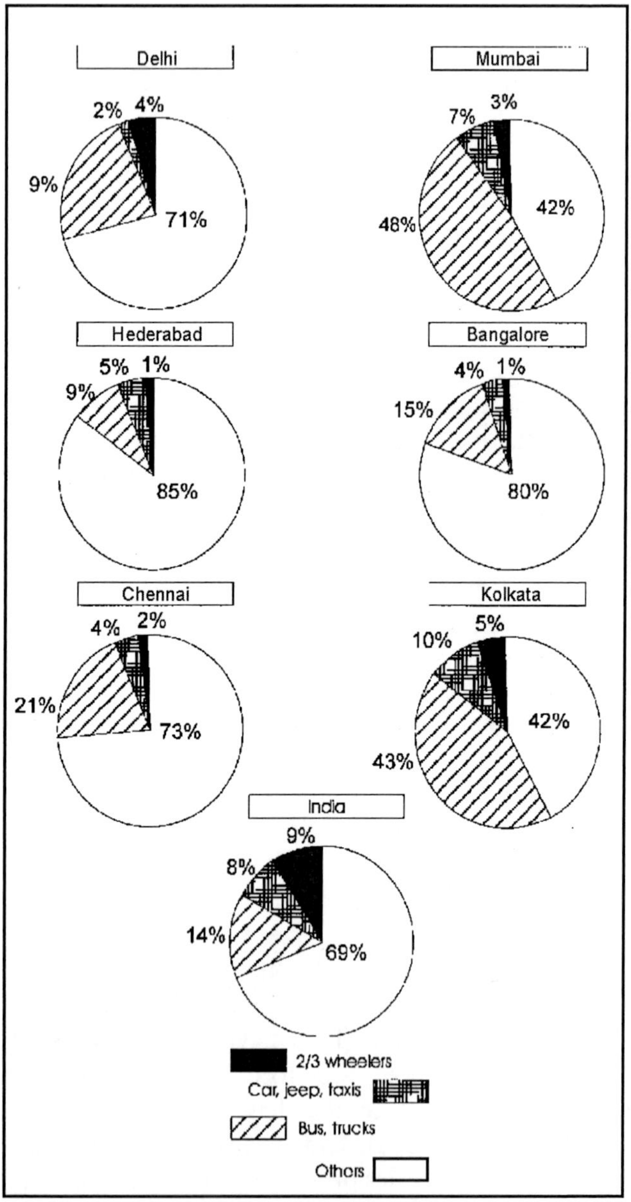

In addition, there is a large concentration of chemical units within 30 km of the city. Similar is the story of other bit cities in India.

Table 16.1

Estimated Vehicular Emission Load in Metropolitan Cities

Name of City	*Suspended Particulate Matter*	*Sulphur dioxide*	*Oxides of Nitrogen*	*Hydro-Carbons*	*Carbon monoxide*	*Total*
Delhi	8.58	7.47	105.38	207.98	542.51	872
Mumbai	4.66	3.36	59.02	90.17	391.6	549
Bangalore	2.18	1.47	21.85	65.42	162.80	254
Kolkata	2.71	3.04	45.58	36.57	156.87	245
Ahmedabad	2.46	2.41	33.33	56.46	149.28	244
Pune	1.99	1.07	13.50	61.0	135.2	213
Chennai	1.95	1.68	23.51	42.05	119.35	188
Hyderabad	1.62	1.30	14.03	46.94	105.14	169
Jaipur	0.98	1.04	12.74	17.49	42.73	75
Kanpur	0.88	0.90	11.14	18.53	40.35	72
Lucknow	0.95	0.79	8.07	18.75	41.02	70
Nagpur	0.46	0.34	4.24	13.60	29.16	48
Grand Total	29.42	24.87	352.39	674.96	1916.01	2998

Coal-based thermal plants, which have been established in the country for power generation, are also responsible for air pollution through gaseous emissions of sulphur dioxide, nitrogen oxide, etc. Air pollution can also cause acid rain, which not only adversely affects man but is also responsible for damage of soil, vegetation and aquatic life of the region and also produces huge amount of solid wastes, fly ash and bottom ash.

Due to air pollution, life in big cities of India has become intolerable due to health problems. Air pollution is the main cause of respiratory illness and loss of vision.

In India, there are several laws to control air pollution, but their implementation is not effective. Pollution by factories is perhaps the easiest to control by imposing standard of emission.

Recently, the Supreme Court has also directed factories in Delhi and its adjoining areas to adopt either proper measures to check pollution or close them. Similar type of action and imposition of heavy penalties is necessary. On the other hand, reducing emission by vehicles is much more complex. A few measures, which can reduce the air pollution, are as follows:

1. Strengthening and introduction of mass transit system as these use significantly less fuel per person per kilometre.
2. Vehicles with old technology should be banned from big cities.
3. Two/three-wheelers should be upgraded to four-stroke engines.
4. Heavy duty vehicles should not be allowed to ply on the inner city roads at peak hours (8 to 12 noon and 4 to 8 p.m.) so as to reduce traffic stagnation.
5. Catalytic converters should be made compulsory for all petrol-driven vehicles and diesel vehicles must have a particle trap.
6. Better urban design to avoid congestion and reduce transport needs.
7. Better and efficient traffic control plan.
8. Encourage the introduction of modern technology in vehicles and also, the use of electric or battery-operated vehicles.

Rural Air Pollution

In India, rural population uses substantial quantities of non-commercial fuel, i.e., crop residues, animal dung or wood in cooking. Although, gradually, its use is decreasing as a percentage of total fuel consumption or in terms of per capita use, but it still provides 80 per cent of rural energy for cooking. Air pollutants due to bio-fuels are largely released directly where people are inside or near households at meal times everyday. Although the emissions are relatively modest, the actual exposure to people is significant in many millions of households around the country. Several health effects are suspected to arise due to smoke such as respiratory infection in children, chronic lung diseases and lung cancer in adults, and adverse pregnancy outcomes, such as low birthweight and still birth, for women exposed during pregnancy.

The rural air pollution can be reduced by: (i) providing an affordable clean fuel for substitute to bio-fuels, (ii) installation of efficient smokeless *chullahs* (much effort has been done in this direction by government), (iii) good ventilation to houses, (iv) providing bio-gas facilities, and above all (v) the habits of rural masses be changed, therefore, awareness regarding ill-effects of the use of bio-fuels should be developed among rural people.

Water Pollution

Once admired for its beautiful water resources, India has over the years progressively depleted its supply of water due to various factors, including increasing population, urbanisation, deforestation and the resultant soil erosion and flawed water management. The available water is also polluted in some way or other. Surface water sources mainly include rivers, springs, lakes and ponds.

The problem of water pollution in India is increasing day by day. The major sources of water pollution in the country are: (i) domestic effluents, (ii) sewage, (iii) industrial effluents, (iv) agricultural effluents, (v) thermal pollution, (vi) oil pollution, and (vii) pollution created by radioactive elements. Pollution from domestic and human waste water causes many severe water-borne diseases. Sewage effluents from cities dotting the river's bank contribute to nearly three-fourths of the water pollution. According to a survey, out of 142 big cities of India, only eight have proper sewage system, 52 have partial system and 82 do not have any sewage disposal system.

River Pollution

River pollution in India has now reached a critical point. Almost every river system in India is now polluted to a great extent. As assessed by the scientists of the National Environmental Engineering Research Institute (NEERI), Nagpur, nearly 70 per cent of river water in India is polluted. India has five major river systems, namely, the Ganga, the Brahmaputra and Indus river systems in north, and the peninsular, east coast and the west coast river systems in the south.

North Indian Rivers

Many large rivers are closely associated with Indian culture and heritage. The pollution situation in our country is worse than that of some of the industrialised countries of Europe and America. The Ganga, the most sacred and important river of India, is regarded as the cradle of Indian civilisation. The 2,525 km long river starts from Gangotri in the Himalayas and joins the Bay of Bengal, at Ganga Sagar. According to a report of the Central Pollution Control Board, despite river Ganga's considerable resilience as a self-purifying and fast-flowing river, its organic pollution load is significantly high. At Kanpur, 45 tanneries and 10 textile mills are the major sources of liquid wastes discharged into the river. The wastes contain heavy organic load and putrefied material. It is estimated that 1,400 million litres of sewage and 200 million litres of industrial effluents are being discharged everyday into the river Ganga. The BOD (106 tonnes/day), total solids (2,308 tonnes/day) and suspended solids (1,251 tonnes/day) are not only very high but also exceed the prescribed limits of the Bureau of Indian Standards. Recently, discharges from the Barauni Oil Refinery caused gross pollution along a long stretch of the main Ganga. Preliminary observations were made on the pollution of the river Kali and a limnological survey was made of the river with reference to fish mortality. The main factories, which pollute the stream are sugar, distillery, tin, glycerin, paints, soap works, spinning, rayon, silk and yarn.

A major step to control and clean the river Ganga had been taken in 1984, when the Central Ganga Authority was established to implement the Ganga Action Plan. This plan has identified 27 cities and about 120 factories as points of pollution from Haridwar to Hooghly. Similarly, the Yamuna Action Plan has also been devised. But till now nothing substantial has come out and there is a long way to go.

A survey was carried out in 1965 along 21 km stretch of the river Gomati in the vicinity of Lucknow receiving 19.84 million gallons of wastes per day from pulp and paper factory, distillery and sewage. Heavy fish mortality was reported in Rihand reservoir due to high free chlorine content (62 ppm) discharged from Kanoria

Chemical Industries. The organic wastes from a sugar factory and a distillery cause year-round pollution in the small river Dhaha. A case of heavy fish mortality was recorded in 1962 and 1966. Wastes from different factories, such as those engaged in the manufacture of paper, chemicals, sugar, cement, etc., are the major sources, discharging over 4 million gallons of wastes per day into the river Sone.

The river Hooghly at Kolkata receives wastes from various types of factories dealing with pulp and paper, distillery, tannery, textile, heavy chemicals, paints and varnishes, shellac, hydrogenated oil, matches, cycle rim, petroleum oil, tar pigment, insecticides and fungicides. Of these, wastes from paper and pulp, distillery, chemicals, textiles, shellacs and a number of domestic outfalls contribute substantially to the pollution complex. River Damodar, which flows through the coal belt area in Bihar, is also a seriously polluted river which is the receptacle for wastes released from large number of industries such as the Sindri unit of the Fertiliser Corporation of India, the Bihar government's superphosphates factory and the associated cement company. The entire Asansol-Durgapur industrial belt on lower Damodar valley suffers from severe pollution caused by the discharge of wastes containing high phenol, cyanide and ammonical nitrogen.

The indiscriminate discharge of large volumes of highly putrefied wastes creates serious pollution problems in the river Ib (Orissa) of the Mahanadi river system. The river Bhadra (Krishna river system) receives effluents from pulp and paper and steel industries.

Industries generate a significant quality of waste water, which ultimately finds its way to a stream/river. Industrial discharges containing toxic and hazardous substances contribute to the severe kind of pollution in the aquatic systems. Industrial development is largely because of the production of chemicals resulting in the generation of toxic and hazardous substances which have been continuously on the increase during the last four decades.

Industrial effluents, though comparatively lesser in volume, cause serious menace to aquatic environment and biotic communities including fish and ultimately affect man through the food chain.

South Indian Rivers

River Godavari is polluted by the effluents of paper mill at Rajahmundry (Andhra Pradesh). The river Kalu in Mumbai receives highly acidic and untreated wastes from Amar Dye and Chemical Company, India Dyes, Century Rayon, National Rayon, Central Chemicals, etc. Effluents of the Gwalior Rayon Factory at Mavoor, about 21 km from Beypore, have created a pollutional hazard in the river Chaliyar at Calicut, Kerala. A large scale fish mortality in 1966 was attributed to highly putrefied organic matter creating almost anaerobic conditions in the river with very low or nil oxygen.

In Tamil Nadu, river Cooum at Chennai gets polluted by the washings from a large number of slums, cattle yards, overflow from sewage pumping station, wastes from automobile workshops and many factories. Nearly 300 tanneries are spread along the banks of river Palar over a stretch of 120 km from Vaniambadi to Ranipet. The waste water discharge from these tanneries affects the ground water quality because of sodium and chlorides present in the tannery waste. River Cauvery is polluted by Mettur Chemical and Industrial Corporation Ltd. River Vaigai receives effluents from many chemical and soap factories and large quantities of municipal sewage.

The rate of pollution in South Indian rivers seems to be higher than those in the North Indian rivers. The main reason for this is the summer season during which the icebergs get melted and the melted water is drained into the major North Indian rivers like the Ganga, the Brahmaputra, etc. During the rainy season rivers get a good current from the natural showers itself. The only possibility of pollution for such rivers is by the catchment of pollutants when they pass through the urban and industrial belts, whereas the South Indian rivers depend mainly on the north-east and the south-west monsoons. Deforestation also affects rainfall and ultimately changes the river flow in South Indian rivers. Already, these rivers are under stress due to lack of rainfall caused by deforestation and pollution is an additional/secondary stress.

Access to safe drinking water remains an urgent need as only 72 per cent of the population in urban area receives well treated water supply. The situation in rural areas is much worse. In India, almost all surface water sources are contaminated and unfit for human consumption. The diseases commonly caused due to contaminated water are diarrhoea, trachoma intestinal worms, hepatitis, etc. Recent WHO data show that about 21 per cent of all communicable diseases in India are water-borne diseases. Epidemiological studies have shown that diarrhoea and intestinal worm infections account for an estimated 10 per cent of the total burden of disease.

An uncontrolled disposal of urban wastes into water bodies, open dumps and poorly designed landfills causes surface water and ground water contamination. Industrial wastes containing heavy metals, such as mercury, chromium, lead and arsenic, can threaten or destroy marine life. Health hazards such as malaria, filariasis, hepatitis and cholera can also spread. The epidemic of jaundice is also a result of contaminated water supply. In 1955–56, in Delhi, about 29,300 persons developed jaundice, following contamination of the drinking water supply from contents of a severed drain from Najafgarh in West Delhi. Similarly, epidemic of hepatitis-E has been caused due to drinking of polluted water. In India, major incidence of hepatitis epidemics has occurred in various urban centres in recent years as shown in Figure 16.2.

A few remedies to control the water pollution are:

- industries should enforce standards for water effluents;
- the Pollution Control Boards must enforce the laws strictly;
- adequate sewage disposal facilities should be developed, so that sewage is not directly released into the water stream;
- the sewage and water treatment plant should be established by every municipal body;
- for small and medium-sized industries, which cannot afford effluent plants, combined treatment plants be established;
- a pollution tax be imposed, especially on industries responsible for water pollution;
- more and more *Sulabh* type toilets be built;
- action plans concerning water pollution should be completed within specified period; and

Figure 16.2

Hepatitis Epidemics in India

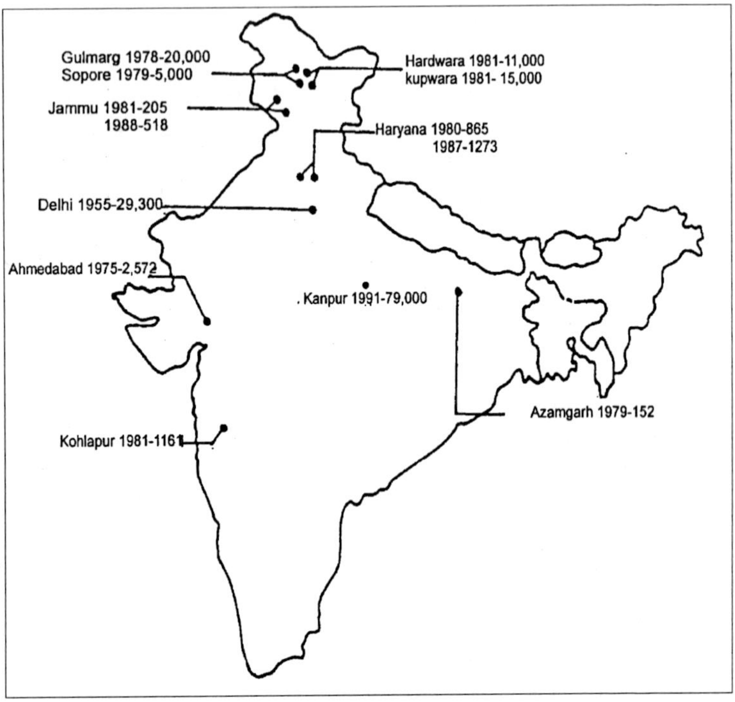

— above all, awareness among people about sanitation and its related impact on their health should be promoted.

Solid Waste Disposal

The solid waste generation is increasing day by day not only in big cities of India but also in small towns and has become a threat to the environment. The composition and quantity of waste changes with increase in per capita income. The main change in solid waste material has been with the increased use of plastic and other synthetic materials in place of biodegradable organic materials. The solid wastes include domestic and municipal waste, industrial and agricultural waste, waste created by mining and also radioactive waste. The composition of solid wastes in five big cities of India has been given in the following table:

Table 16.2
Composition of Solid Wastes from Cities

Cities	*Characteristics (%)*					
	Non-degradable					*Degra-dable*
	Paper	*Plastics*	*Metal*	*Glass*	*Ash and Earth*	
Kolkata	3.18	0.65	0.66	0.38	34.00	47.00
Delhi	6.29	0.85	1.21	0.57	36.00	35.00
Nagpur	1.88	1.35	1.33	1.34	41.42	34.81
Bangalore	4.00	2.00	—	1.00	15.00	78.00
Mumbai	10.00	2.00	3.6	0.2	44.20	40.00

When this solid waste is not collected and disposed of efficiently and effectively, it attracts rodents and flies which then spread diseases and also pollute water and cause land degradation. The status of solid waste collection in selected cities is shown in Table 16.3.

Table 16.3
Collection of Solid Wastes in Cities and Annual per Capita Solid Waste Generated

Cities	*Solid Waste*	
	Generated (Mt)	*Collected (%)*
Mumbai	0.20	90
Delhi	0.44	77
Chennai	0.29	90
Bangalore	0.16	96
Lucknow	0.31	74
Varanasi	0.14	87
Hubli-Dharwad	0.15	89
Mysore	0.26	60
Gulbarga	0.10	40
Tumkur	0.13	50

These wastes can be utilised after sorting for generating energy and useful resources by way of composing, vermicompost and anaerobic digestion. The potential of power generation from these wastes is tremendous. It is estimated that urban municipal wastes have a potential to generate upto 100 MW of electricity. Industrial wastes coming from dairies, distilleries, perfumes, tanneries, pulp and paper and food processing industries have the potential to generate about 700 MW of power.

These wastes are left untreated, they ferment slowly and produce biogas that contains 65–75 per cent methane gas, which is a greenhouse gas. Therefore, development of suitable technologies for utilisation of wastes is essential to minimise adverse health and environmental consequences.

Deforestation

Forests have a many-faced ecological role to play which affects human life in a variety of ways. They thwart the dangers of cloud drifting, soil erosion, floods, wind erosion and ground water evaporation. They also protect a wide variety of flora and fauna, provide recreation and can effectively control air pollution of moderate magnitudes. But, due to industrialisation, commercialisation, growth of population and expansion of agriculture, in India, forest cover is fast disappearing. Table 16.4 indicates the state-wise forest cover in India.

India had forest area of 6,75,538 sq km in 2005, which constitutes only 20.55 per cent of the total land area. A large part of these forests is degraded and productivity is very poor. Even in protected forests unrecorded changes are going on. Actually, the closed type of forests having forest cover density of 40 per cent or more is approximately 11 per cent of the country's total land area. Forest cover situation as assessed by remote sensing, however, shows encouraging results of an increase in the area under closed forests. The comparative situation of forest cover in 1991 and 2001 has been given in Table 16.4.

Table 16.4
Forest Cover in 1991 and 2001 (in thousand sq km)

State/UTs	*1991*	*% of Total Area*	*2001*	*% of Total Area*
Andhra Pradesh	47.3	17.2	44.6	16.23
Arunachal Pradesh	68.7	82.1	68.0	81.2
Assam	24.7	31.6	27.7	35.3
Bihar	26.6	15.3	5.7	6.0
Delhi	0.002	1.5	0.10	7.5
Goa, Daman and Diu	1.2	33.8	2.0	56.6
Gujarat	11.9	6.1	15.1	7.7
Haryana	0.5	1.2	1.7	3.9
Himachal Pradesh	11.7	21.2	14.3	25.8
Jammu and Kashmir	20.0	9.0	2.2	9.5
Karnataka	32.2	16.8	36.9	19.3
Kerala	10.3	26.5	15.5	40.0
Madhya Pradesh	135.6	30.6	77.21	25.0
Maharashtra	44.6	14.3	47.4	15.4
Manipur	17.7	79.2	16.9	75.8
Meghalaya	15.9	70.8	15.5	69.5
Mizoram	18.9	89.4	17.4	82.9
Nagaland	14.3	86.4	13.3	80.5
Orissa	47.2	30.3	48.8	31.4
Punjab	1.3	2.7	2.4	4.8
Rajasthan	12.8	3.7	16.3	4.78
Sikkim	3.0	42.8	3.1	45.0
Tamil Nadu	17.7	13.6	21.4	16.5
Tripura	5.5	52.8	7.0	67.4
Uttar Pradesh	33.6	11.4	13.7	5.7
West Bengal	8.0	9.0	10.6	12.5
A&N Islands	7.6	92.4	6.9	84.0
Chandigarh	.005	4.4	0.01	7.9
Dadra and Nagar Haveli	.206	42.0	0.21	44.6
Lakshadweep*	—	—	—	—
Pondicherry*	—	—	—	—
Total	639.2	19.44	675.5	20.05

* No discernible forest cover.

Source: The State of Forest Report 1995 and 2001, Ministry of Environment and Forests.

A wide range of flora and fauna are fast disappearing as their natural habitats are destroyed. There is thus need for massive reforestation programmes. The effective afforestation, reforestation and regeneration activities should be carried out. Agro-forestry should also be encouraged. For rural population cheap and clean fuel should be provided to save the forests from denudation.

Soil Degradation

In India, about 130 million hectares of land (45% of total geographical area) is affected by serious soil erosion through ravine and gully, shifting cultivation, cultivated wastelands, sandy areas, deserts and waterlogging (Government of India, 1989). The soil formation is a very long process and under favourable conditions, in India, it takes about thousand years to form only 2.3 cm of a soil layer. It is reported that the loss of 1 mm of cultivated soil could cost 10 kg of nitrogen and 2 kg of phosphorus. Soil erosion by rains and rivers in hilly areas causes landslides and floods, while cutting trees for firewood and timber, grazing by livestock, traditional agricultural practices, construction of roads, indiscriminate quarrying and other activities also lead to the opening of soil erosion.

The main agents of soil erosion in India are water, wind, shifting cultivation and sea-waves. Water erosion is a main cause of soil erosion in the form of gully erosion in Chambal Valley in Madhya Pradesh, Rajasthan and Uttar Pradesh as well as in other river valleys of the country. On the other hand soil erosion by wind is significant in arid and semi-arid areas of Rajasthan, Haryana, Punjab, Madya Pradesh and Gujarat.

The causes of the degradation of soil in India varies from region to region. The Planning Commission, has mentioned the types of degraded soils/land and their causes. These categories of land and their causes have been given in the Table 16.5.

Table 16.5
Type of Degraded Land

Categories	*Area in Million Hectares*
Gullied Lands	2.05
Land with/without scrub	19.40
Waterlogged and marshy lands	1.66
Land affected by salinity and alkalinity	2.05
Shifting cultivation area	3.51
Unutilized/degraded forests	14.06
Degraded pastures and grazing of lands	2.60
Degraded land under plantation	0.58
Sand (inland/coastal) crops	5.00
Mining/industrial wasteland	0.12
Barren rock area	6.42
Steep sloping area	0.77
Snow covered/glacial area	5.58
Total	63.84

Biodiversity

India has a rich heritage of species and genetic strains of flora and fauna. Overall, 8 per cent of world species are found in India. It is estimated that India ranks tenth among the plant-rich countries of the world. It has diversity in its ecosystem consisting of forests, wetlands, grassland, deserts, marine areas, coral reefs, etc., and each system supports thousands of species. The total number of living species identified in India so far is 2,00,000 (many more not yet discovered). The pattern of biodiversity in India is presented in Table 16.6.

But India's rich biological diversity is rapidly eroding. More than half of its forests are gone, two-thirds of its wetlands drained out or polluted, and almost all of its dry scrub and grassland habitats destroyed. Habitat destruction, hunting and overexploitation, pollution and poisoning, and introduction of exotics have caused a decline in many species. Many species have gone extinct, of what remains, over 10 per cent of flowering plants, 21 per cent mammals and 5 per cent of birds are believed to be under threat.

Table 16.6
Biodiversity of India

Species	*Number*
Total plants	45,000
Mammals	340
Reptiles	420
Amphibians	2000
Fishes	1693
Insects	67000
Molluscs	4000
Birds	1200
Invertebrates	6800

Source: Negi, S.S. (1991), *Environmental Degradation and Crisis in India*, Indus Pub., New Delhi, pp. 68–69.

At the Convention on Biological Diversity (CBD), signed at the Earth Summit in 1992, India was the first country to sign it. Not only this, but several efforts have also been made to conserve the rich heritage of biodiversity. In order to preserve it, 358 wildlife sanctuaries and 63 national parks have been developed in India. There is thus need to implement the laws of the protection of forest and wildlife. What is needed more is the people's participation. Two measures, which might greatly help, are:

1. Local communities must be given more secure and meaningful rights over public lands and waters in their own localities.
2. Local communities should be given substantial incentives to continue conserving biodiversity resources on private and public lands in their own localities by linking financial inputs.

Natural Hazards and Their Impact on Environmental in India

In every part of the world, several natural phenomena, such as volcanic eruptions, earthquakes, cyclones, landslides, floods, droughts and famines, have had their impact on environment. Sometimes their impact is sudden and normal but often it is adverse not only to man but also to the environment and therefore are termed as natural hazards, environmental hazards, environmental stresses, environmental disasters, extreme events, etc. These natural hazards are common in various parts of India.

Since India is not located in the volcanic zone, volcanic eruption is not common. But, in November-December 1991, at Andaman Island, a dormant volcano became active and caused damage to vegetation, marine life and also polluted the atmosphere. The incidence of landslide is common in hilly areas of Jammu & Kashmir, Himachal Pradesh and Uttrakhand. Similarly, coastal regions are susceptible to cyclonic storms, which cause massive destruction of life and property as well as to vegetation. The coastal regions of Orissa, West Bengal, Andhra Pradesh and Tamil Nadu often face the problem of cyclones.

In India, earthquake-affected areas include Jammu & Kashmir, Himachal Pradesh, West Bengal and the north-eastern states. These are located in the foothills of Himalayas, while there are other states, such as Punjab, Haryana, Delhi, Rajasthan and Maharashtra, which are also affected by earthquakes. The impact of earthquakes on environment is much more in the form of destruction of flora and fauna, occurrence of rifts and cracks in land, change in course of rivers, cracks in dams and a great loss of man and property, depending upon the intensity of the quake. On 11 October 1937, an earthquake claimed the lives of over 3 lakh people. Other important earthquakes have occurred on 30 May 1885 in Kashmir, 12 June 1887 at Shillong, 15 June 1870 in Kacch, 4 April 1905 in Kangra Valley, 5 January 1934 in Bihar, 15 August 1995 in Assam, 11 December 1967 at Koyana (Maharashtra), 21 August 1988 at Darbhanga (Bihar), 20 October 1991 in Uttar Kashi, 30 September 1993 in Maharashtra, etc.

In India, occurrence of floods during the monsoon season is very common. Floods are very common in states like Bihar, Uttar Pradesh, Uttrakhand, Andhra Pradesh, West Bengal, Assam,

Gujarat, Tamil Nadu, Haryana, and Madhya Pradesh. In India, the total area affected by floods is about 1.37 million hectares, while the total affected crop area 157.6 million hectares. Effective flood control measures, such as utilisation of flood water by construction of small dams, afforestation in catchment area and construction of channels, etc., can minimise its hazardous effect.

While a good monsoon becomes a cause of floods in India, the failure of monsoon is responsible for drought resulting in famines. Droughts in turn become cause of environmental degradation. The famine that occurred in year 1899, known as *Chappania Akkal*, is well known in the history of India, because of its widespread impact and death of several lakhs of people. A similar famine occurred in 1917. In 1986–87, about two-third of the country was under the grip of famine. The worst hit area is Rajasthan, where famine often occurs due to desertic conditions and failure of monsoon. During the last 20 years, Rajasthan has had famine in 16 years. The development of irrigation facilities and proper water management can help solve this problem.

Environmental Planning and Management in India

In earlier chapters as well as in this chapter, references have been made regarding various efforts done in India for the protection of the environment. In fact, India is a country where all the components of environment have been considered as life support systems right from ancient times and their protection has been of paramount importance. The air, water, land, vegetation, soil, rivers, mountain, trees, animals, etc., have been considered as sacred in one form or the other. There was a general public opinion to protect all these as a part of religious activity. Although at that time there was no problem of environment because of limited population and simple pattern of life, gradually, especially during the last century, with the growth of population, industrialisation, urbanisation as well as growth in other economic activities, environmental problems have developed, therefore, a planning strategy is necessary.

For the protection of environment and also to control pollution, several efforts have been undertaken, both by government and NGOs. Even before independence, some laws have been enacted for the protection of the environment. In Indian Penal Code of 1860, Articles 268, 290, 291, 426, 430, 431 and 432 are related with the environment. Similarly, Article 277 was related with water pollution and Article 278 with air pollution. In Motor Vehicle Act, 1938, there was a provision to control pollution and Indian Forest Act was passed in 1927. After independence, serious efforts have been made in the form of legislation for the conservation and protection of the environment, the Act, and laws passed during last sixty years, is a major step for the protection of environment in India. The following are the important Acts:

1. Damodar Valley Corporation (Prevention of Pollution of Water) Regulation Act, 1948.
2. River Board Act, 1956.
3. Water Preservation and Control of Pollution Act, 1974 and 1977.
4. Atomic Energy Act, 1972.
5. Radiation Protection Rules, 1971.
6. Wild Life Protection Act, 1972.
7. Factories Amendment Act, 1987.
8. Environmental Protection Act, 1986.
9. Central Motor Vehicles Rules, 1989.
10. Public Liability Insurance Act, 1991.
11. Air (Prevention and Control of Pollution Act), 1981).
12. Noise Pollution (Regulation and Control) Rules, 2000.
13. Hazardous Wastes (Management and Handling) Rules, 1989
14. Bio-Medical Wastes (Management and Handling) Rules, 1989
15. Municipal Solid Wastes (Management and Handling), Rules, 2000.

Besides, in 1977, a major step in this direction has been taken in the form of 42nd Constitution Amendment. Accordingly, Article 48-A imposes a duty on the state "to protect and improve the environment and to safeguard the forests and wildlife of the country". Likewise, Article 51A(g) imposes a duty on citizens of

India "to protect and improve the natural environment including forests, lakes, rivers and wildlife and to have compassion for living creatures".

In India, the Union Ministry of Environment and Forests (MEF) is primarily concerned with the implementation of policies and programmes relating to conservation of the country's natural resources including lakes and rivers, its biodiversity, forests and wildlife, ensuring the welfare of its animals and prevention and abatement of pollution. While implementing these policies and programmes, the MEF is guided by the principle of sustainable development and enhancement of human well-being. The MEF also serves as the nodal agency in the country for the United Nations Environment Programme (UNEP), South Asia Cooperative Environment Programme (SACEP), International Centre for Integrated Mountain Development (ICIMOD) and for the follow-up of the United Nations' Conference on Environment and Development (UNCED). The Ministry is also entrusted with the issues relating to multilateral bodies, such as the Commission on Sustainable Development (CSD), Global Environment Facility (GEF) and of regional bodies like Economic and Social Council for Asia and Pacific (ESCAP) and South Asian Association for Regional Cooperation (SAARC), on matters pertaining to environment.

Conservation and survey of flora, fauna, forests and wildlife, prevention and control of pollution, afforestation and regeneration of degraded areas and protection of environment are the mandates of the MEF. These objectives are sought to be fulfilled through environmental impact assessment, eco-regeneration, assistance to organisations in implementing environmental and forestry programmes, including animal welfare programmes, promotion of environmental and forestry research, education and training, dessemination of environmental information and international cooperation and creation of environmental awareness. A brief review of major steps taken by the government is as follows:

Survey of Natural Resources

The Botanical Survey of India (BSI) was established in 1890 with the basic objective of carrying out floristic surveys. It was revived and reorganised in 1954. During the successive plan periods, its functions have been gradually expanded. A thorough review of the objectives and perspectives was undertaken in 2002. The BSI undertakes exploration, inventorisation and documentation of phyto-diversity in general and protected areas, hotspots, fragile ecosystems and sacred groves in particular, and also publication of National, State and District Floras.

The Zoological Survey of India (ZSI), a premier institute under the Ministry of Environment and Forests, has been undertaking survey, exploration and research leading to the advancement of knowledge on the exceptionally rich faunal diversity of the country since its inception in 1916. With its headquarters at Kolkata and 16 regional stations located in different parts of the country, the ZSI, in recent years, has reoriented its plan of work by grouping the surveys and studies under five major programmes, viz., (i) fauna of states; (ii) fauna of conservation areas; (iii) fauna of important ecosystems; (iv) status survey of endangered species; and (v) ecological studies. Besides the ZSI also undertakes regular faunistic survey tours of the entire country. Eighty-two extensive faunal surveys were undertaken in different states/union territories including important ecosystems and some selected conservation/protected areas during 2003–04.

The Forest Survey of India (FSI), an organisation of Ministry of Environment and Forests, is engaged in generating information and database on forest cover and forest resources in the country besides providing training, research and extension service. It was established in 1981 as a successor to "Pre-investment Survey of Forest Resources" (PISFR), a project initiated in 1965 by Government of India and sponsored by FAO and UNDP. The main objective of PISFR was to ascertain the availability of raw material for establishment of wood-based industries in selected areas of the country. Further, the National Commission on Agriculture (NCA), in its report in 1976, recommended the creation of a National Forest Survey Organisation (NFSO) for collection of data on scientific lines through country-wide comprehensive forest resources surveys at regular intervals. Consequently, PISFR was reorganised into FSI. After a critical review of activities undertaken by FSI, the government, in 1986, redefined its mandate in order to make it more purposeful and relevant to the needs of the country.

Biosphere Reserves

Biosphere reserves are areas of terrestrial and coastal ecosystems, which are internationally recognised within the framework of UNESCO's Man and Biosphere (MAB) Programme. These reserves are required to meet a minimal set of criteria and adhere to a minimal set of conditions before being admitted to the World Network of Biosphere Reserves. The world's major ecosystem types and landscapes are represented in this Network. The goal is to facilitate conservation of representative landscapes and their immense biological diversity and cultural heritage, foster economic and human development, which is culturally and ecologically sustainable, and to provide support for research, monitoring, education and information exchange. So far 13 biosphere reserves have been set up. They are Nilgiri, Nanda Devi, Nokrek, Great Nicobar, Gulf of Manner, Manas, Sunderbans, Similipal, Dibru Daikhowa, Dehong Deband, Panchmarhi, Kanchanjanga and Agasthyamalai. Out of these, three have been recognised on World Network of Biosphere Reserves by UNESCO, namely, Nilgiri, Sunderbans and Gulf of Mannar.

Biodiversity

Biodiversity is the variability among living organisms and the ecological complexes of which they are part, including diversity within and between species and ecosystems. Biodiversity has direct consumptive value in food, agriculture, medicine and industry. India is one of the 17 mega diverse countries, which together possess 60 to 70 per cent of the world's biodiversity.

India ratified the International Convention on Biodiversity (CBD) on 18 February 1994 and became party to it in May 1994. The CBD is an international legal instrument for promoting conservation and sustainable use of biological diversity taking into account the need to share cost and benefit between developed and developing countries and the ways and means to support innovation by local people. It was resolved to evolve an international regime on access to genetic resources and benefit sharing with the aim of adopting an instrument/instruments to effectively implement the provisions of CBD.

The National Policy and Action Strategy on Biodiversity, which seeks to consolidate the ongoing efforts of conservation and sustainable use of biological diversity and to establish a policy and programme regime for the purpose, was released by the MEF on 6 January 2000. The Bio-Diversity Bill, introduced in parliament in May 2000, was passed on 11 December 2002. The main intent of the legislation is to protect India's rich biodiversity and associated knowledge against their use by foreign individuals and organisations without sharing the benefits arising out of such use, and to check biopiracy.

Forests

Forests are a renewable source and contribute substantially to a economic development. They also play a major role in enhancing the quality of environment. The forest cover in India is 6,75,538 sq km and constitutes 20.55 per cent of its geographical area. Of this, dense forests constitute 4,16,809 sq km (12.68%) and open forests 2,58,729 sq km (7.87%). A comparison of forest cover assessment of 2001 with that of 1999 reveals that there is an overall increase of 38,245 sq km or 6 per cent. This constitutes an increase of 1.16 per cent of geographical area. The increase in dense forest cover with respect to 1999 assessment is 34,580 sq km (9%) and increase in open forest cover is 3,665 sq km (1.4%).

Forest Policy and Law

India is one of the few countries having a forest policy since 1894. It was revised in 1952 and again in 1988. The main plank of the forest policy is protection, conservation and development of forests. Its aims are:

(i) maintenance of environmental stability through preservation and restoration of ecological balance;
(ii) conservation of natural heritage;

(iii) check on soil erosion and denudation in catchment area of rivers, lakes and reservoirs;
(iv) check on extension of sand dunes in desert areas of Rajasthan and along coastal tracts;
(v) substantial increase in forest tree cover through massive afforestation and social forestry programmes;
(vi) steps to meet requirements of fuelwood, fodder, minor forest produce and soil timber of rural and tribal populations;
(vii) increase in productivity of forest to meet the national need;
(viii) encouragement to efficient utilisation of forest produce and optimum substitution of wood; and
(ix) steps to create massive people's movement with involvement of women to achieve the objectives and minimise pressure on existing forest.

As forestry has undergone many conceptual changes since the adoption of Indian Forest Act, 1927, it was decided to bring suitable amendments to this. The MEF has constituted the National Forest Commission on 7 February 2003 to review the working of forests and wildlife sector. A National Forestry Action Programme (NFAP) has also been formulated as a comprehensive strategic long-term plan for the next 20 years. The objective is to bring one-third of the area of the country under tree/forest cover and to arrest deforestation.

In 1990, the Government of India issued guidelines to involve the village communities in the development and protection of degraded forests on the basis of their taking a share of the usufruct from such areas. The concept of Joint Forest Management (JFM) was accordingly initiated. The JFM Programme was pursued vigorously and as a result JFM resolution has now been adopted in all the 28 States. 84,632 JFM committees have been formed and Forest area worth 17.33 million hectares has been brought under the JFM programme. About 85.28 lakh families are involved in the programme all over the country.

An Integrated Forest Protection Scheme has been formulated by merging the two schemes, viz., 'Forest Fire Control and Management' and 'Bridging of Infrastructure Gaps in the Forestry Sector in the North-Eastern Region and Sikkim'. The cent per cent centrally-sponsored scheme is now being extended to all the states and UTs during the Tenth Five-Year Plan.

Wildlife

The National Wildlife Action Plan provides the framework of strategy as well as programme for conservation of wildlife. The first National Wildlife Action Plan (NWAP) of 1983 has been revised and the new Wildlife Action Plan (2002–2016) has been adopted. The Indian Board of Wildlife, headed by the Prime Minister, is the apex advisory body overseeing and guiding the implementation of various schemes for wildlife conservation. At present, the protected area network comprises 92 national parks and 500 sanctuaries covering an area of 15.67 million hectares. The Wildlife (Protection) Act, 1972, adopted by all states except Jammu & Kashmir (which has its own Act), governs wildlife conservation and protection of endangered species. It prohibits trade in rare and endangered species. An Inter-State Committee has been constituted to review the Wildlife (Protection) Act, 1972 and other laws. India is a signatory to the Convention on International Trade in Endangered Species (CITES) of wild flora and fauna. It is also a signatory to the Memorandum of Understanding concerning the conservation of Siberian cranes. The government provides financial and technical assistance to states for development and improved management of national parks, protection of wildlife and control of poaching and illegal trade in wildlife products, eco-development in areas around national parks and sanctuaries, conservation of elephant and its habitat, and conservation of rhinos in Assam.

Under the Project Tiger, launched on 1 April 1973, 281 Tiger Reserves have been set up in 17 states covering an area of about 37,761 sq km. While under the Project Elephant, which was launched in February 1992, states having free-ranging population of wild elephants are being given financial as well as technical and scientific assistance to ensure long-term survival of identified viable populations of elephants in their natural habitats. As many as 14 Elephant Reserves have been created during the year. A Central Zoo Authority has also been formed to look after the management of zoological parks in the country. It coordinates the activities of animals on a scientific basis. A national policy on zoos provides appropriate directions to the government and other zoo operators.

Environment Impact Assessment

For sustainable development and optimal use of natural resources, environmental considerations are required to be integrated in planning, designing and implementation of development projects. Environmental Impact Assessment (EIA) is one of the proven management tools for incorporating environmental concerns in development process and also in improved decision-making. The programme of EIA, in vogue in the MEF for the last two decades, was initiated with the appraisal of river valley projects. The scope of appraisal was subsequently enlarged to cover other sectors like industrial projects, thermal power plants, mining schemes and infrastructure projects. To give legislative status to the procedure of impact assessment, EIA was made mandatory since January 1994 for 30 categories of development activities. For facilitating preparation of quality EIA reports, a Manual on EIA has been prepared, which is also useful to apprising agencies and decision-makers, both at central and state levels.

Mining proposal involving major minerals with lease area more than 5 hectares in areas covered by the Aravalli notification has been brought under the purview of EIA notification. Widening and strengthening of existing canals with land acquisition up to a maximum of 20 metres (on both sides put together) along the existing alignments, provided such canals do not pass through ecologically sensitive areas, has been exempted from the purview of EIA notification.

The distance criterion, which was 25 km in case of reserved forests and ecologically sensitive areas, and 50 km in case of interstate boundary, has been reduced to 15 km in both the cases. As a result, power projects of co-generation plants, captive power plants up to 250 MW (both coal and gas/naptha based) coal-based plants up to 500 MW using fluidised bed technology, coal-based power plants up to 250 MW using conventional technology, gas/naptha-based plants up to 500 MW and not falling within a critically polluted area, or within a radius of 15 km of boundary of reserved forests, ecologically sensitive areas in any state, can be considered and accorded environmental clearance by the state governments themselves.

With the environmental site clearance made mandatory for mega projects, such as green field airports, petro-chemical complexes and refineries, isolated petroleum product storages have been included along with petroleum refineries including crude and product pipelines. Public hearing has been exempted for offshore exploration activities, beyond 10 km from the nearest habitated village boundary, goothans and ecologically sensitive areas such as mangroves (with a minimum areas of 1,000 sq mt) coral reefs, national parks, marine parks, sanctuaries, reserve forests and breeding and spawning grounds of fish and other marine life.

In order to improve the environmental clearance process and to make it more effective and time-bound, the Ministry of Environment and Forests has undertaken a review of environmental clearance under the World Bank-assisted Environmental Management Capacity Building (EMCB) project.

For the purpose of protecting and conserving the coastal environment, the MEF had issued the Coastal Regulation Zone (CRZ) notification, on 19 February 1991, declaring coastal stretches of seas, bays, estuaries, creeks, rivers and back waters, which are influenced by tidal action (on the land ward side), up to 500 metres from the high tide line and the inter-tidal zone, as the CRZ. The notification imposes restrictions on the setting up and expansion of industries and operations or process, etc., in the CRZ.

Prevention and Control of Pollution

Realising the deteriorating air and water quality, increasing vehicular emission and higher noise levels, the MEF formulated a policy for abatement of pollution, which provides multi-pronged strategies in the form of regulations, legislations, agreements, fiscal incentives and other measures. Over the years, the thrust has shifted from curative measures to prevention and control through adoption of clean and low-waste technology, reuse and recycling, natural resource accounting, environmental audit and human resource development. To give effect to environmental measures and policies for pollution control, various steps have been initiated, which include stringent regulations, development of environmental standards, control of vehicular pollution, spatial environmental planning including industrial estates and preparation of zoning atlas. The scheme on development and promotion of clean technologies, therefore, aims at promoting such technologies and strategies.

An "*Eco-mark*" lable has been introduced to put on consumer products that are environment-friendly. So far, the government has issued 19 notifications on different products criteria. Submission of an environmental statement by polluting units seeking consent either under the Water (Prevention and Control of Pollution) Act, 1974 or the Air (Prevention and Control of Pollution) Act, 1981 or both and the authorisation under the Hazardous Wastes (Management and Handling) Rules, 1989 has been made mandatory through a gazette notification in April 1993 under the Environment (Protection) Act, 1986. The primary benefit of environmental audit is that it ensures cost-effective compliance of laws, standards, regulations and company policies, etc.

The policy statement for abatement of pollution lays emphasis on preventive aspects of pollution abatement and promotion of technologies to reduce pollution. As a part of the industrial pollution abatement through preventive strategies, financial assistance is being provided for establishment and running of waste minimisation circles in clusters of small-scale industries, capacity building in areas of cleaner production, establishment of demonstration units in selected industrial sectors, etc. The programme is being funded through National Productivity Council, which has pioneered in the activity.

A network of 295 Ambient Air Quality Monitoring Stations, covering 90 towns/cities all over the country, has been set up by Central Pollution Control Board in coordination with the State Pollution Control Boards/Pollution Control Committees and other institutions for carrying out regular monitoring. The major objectives of the programme are: (i) to ascertain whether the notified ambient air quality standards are maintained; (ii) to control and regulate pollution from various sources; (iii) to understand the natural cleansing process in the environment through pollution dilution, dispersion, wind-based movement, dry deposition, precipitation and chemical transformation of pollutants generated; and (iv) health impacts.

Under National Air Quality Monitoring Programme (NAMP), four air pollutants, viz., sulphur dioxide (SO_2), oxides of nitrogen as NO_2, Suspended Particulate Matter (SPM) and Respirable Suspended Particulate Matter (RSPM/PM_{10}), have been identified for regular monitoring at all the locations. Besides this, additional parameters such as respirable lead and other toxic trace metals and polycyclic aromatic hydrocarbons (PAHs) are also being monitored in seven metro cities of the country. The monitoring of meteorological parameters, such as wind speed and direction, relative humidity and temperature, was also integrated with the monitoring of air quality. The monitoring of pollutants is carried out for 24 hours with a frequency of twice a week, to have 104 observations in a year.

As vehicular emission is the major cause for deterioration of urban ambient air quality, The MEF is facilitating and coordinating controlling of vehicular pollution in the field with the concerned ministries and its associated bodies/organisations including the Ministry of Surface Transport, the Ministry of Petroleum and Natural Gas and the Ministry of Industry, in areas such as upgradation of automobile technology, improvement in fuel quality, expansion of urban public transport systems and promotion of integrated traffic management, etc. The gross emission standards for vehicles have been prescribed from time to time and a road map is prepared to improve the quality of fuel.

The Central Pollution Control Board (CPCB), an autonomous body of the MEF, was set up in September 1974, under the provisions of the Water (Prevention and Control of Pollution) Act, 1974. It coordinates the activities of the State Pollution Control Boards (SPCBs) and Pollution Control Committees (PCCs), and also advises the Central Government on all matters concerning the prevention and control of environmental pollution. CPCB, SPCBs and PCCs are responsible for implementing the legislation, regulations and guidelines relating to prevention and control of pollution;

they also develop rules and regulations which prescribe the standards for emissions and effluents of air and water pollutants and noise levels. CPCB also provides technical services to the MEF for implementing the provisions of the Environment (Protection) Act, 1986.

Management of Hazardous Substances

Hazardous Substances Management Division is the nodal point within the MEF for planning and overseeing the implementation of policies and programmes on management of hazardous substances and chemical emergencies. The objective is to promote safe handling, management and use of hazardous substances, including hazardous chemicals and hazardous wastes, in order to prevent potential damage to health and environment. The activities are carried out under three thrust areas, i.e., chemical safety, sound management of hazardous wastes and municipal solid wastes. Various rules have been notified to ensure environmentally sound management of hazardous substances and wastes in the country.

A Central Control Room has been set up in the MEF to deal with emergencies arising from hazardous chemicals. A Crisis Alert System has also been established. A Red Book entitled "Central Crisis Group Alert System" containing names, addresses and telephone numbers of central and state authorities and experts to be contacted in case of chemical emergencies has been published and is regularly updated. A comprehensive National Chemical Profile is being prepared. Presently there are 1,580 Major Accident Hazard (MAH) units in 234 districts of 19 states/UTs of the country. Financial assistance is being provided to the state governments to strengthen infrastructure for preparation of crisis management plans. So far, 1,107 on-site plans and 138 off-site plans have been prepared. Most states have constituted State Level Crisis Groups. Under the Public Liability Insurance Act, 1991, as amended in 1992, all the MAH units handling chemicals in excess of the threshold quantities referred to in the Schedule are mandated to take an insurance policy and deposit an equal amount in the Environment Relief Fund to ensure immediate payment to chemical accident victims. Out of 180 hazard-prone industrial pockets, hazard analysis studies have been initiated for 75 pockets.

As per current assessment 4.4 million tonnes of hazardous wastes are being generated by 13,011 units spread over 373 districts in the country. The states of Maharashtra, Gujarat and Tamil Nadu account for over 63 per cent of the total hazardous wastes generated in the country. This data, which is based on the waste categories indicated in the Hazardous Wastes (Management and Handling) Rules, 1989, is being revised in the light of the amendments carried out in 2000 and 2003 and also in view of the directions of the Supreme Court.

The Municipal Solid Wastes (Management and Handling) Rules, 2000, Fly-ash Notification, 1999 and the Recycled Plastics (Manufacture and Usage) Rules, 1999 (amended in 2003) constitute the regulatory framework for the management of solid wastes in the country. Use of fly-ash in construction, laying of roads and reclamation of low-lying areas has been made mandatory. Guidelines for use of fly-ash have also been formulated and circulated to the state Governments. Manufacture and usage of plastic carry bags less than 8x20 inches in size has been banned.

The MEF is also the nodal point for three international conventions, namely, the Basel Convention on the Control of Trans-boundary Movement of Hazardous Wastes and Their Disposal, the Rotterdam Convention on the Prior Informed Consent (PIC) Procedure for Certain Hazardous Chemicals and Pesticides in International Trade, and the Stockholm Convention on Persistent Organic Pollutants (POPs). As a signatory to the Basel Convention, the MEF participates in all the technical and legal working group meetings. Though India signed the Stockholm Convention on POPs on 14 May 2002, it is yet to ratify. It seeks to eliminate production use, import and export of 12 POPs wherever techno-economically feasible, and in the interim period, restrict the production and use of these chemicals.

National River Conservation Directorate

The National River Conservation Directorate (NRCD), functioning under the MEF is engaged in implementing the River Action Plan under the National River Conservation Plan (NRCP) by providing assistance to the state governments. The objective of

NRCP is to improve the water quality of the rivers, which are the major fresh water sources in the country, through the implementation of pollution abatement schemes. So far, a total of 31 rivers have been covered under the programme.

The first River Action Plan to be taken up under the NRCD was the Ganga Action Plan (GAP). The objective of the GAP Phase-I was to improve the water quality of the river Ganga. It primarily addressed itself to the interception and diversion for treatment of 873 million litres per day (MLD) of municipal sewage in 25 towns. GAP (Phase-I) scheme has been closed on 31 March 2000. A sewage treatment capacity of 865 MLD has been created under the programme and an amount of about Rs 452 crore has been spent. The water quality of the river Ganga is being monitored at 27 locations from Rishikesh in Uttaranchal to Uluberia in West Bengal.

The Yamuna Action Plan (YAP) presently covers pollution abatement works in 21 towns. The approved cost of YAP is Rs 509.45 crore, of which an expenditure by the states of Rs 474.57 crore has been incurred so far. A treatment capacity of 738 MLD has been created. Of the 179 schemes of pollution abatement sanctioned under YAP, 161 schemes have been completed. Additional pollution abatement works in 15 towns of short gestation period amounting to Rs 222.60 crore were approved in May 2001. All the works have been completed under this package.

National Afforestation and Eco-development Board

The National Afforestation and Eco-development Board (NAEB) was set up in August 1992 for promoting afforestation, tree planting, ecological restoration and eco-development activities in the country. Special attention is being given to regeneration of degraded forest areas and lands adjoining forest areas, national parks, sanctuaries and other protected areas as well as the ecologically fragile areas like the Western Himalayas, Aravallis, Western Ghats, etc.

The NAEB has evolved specific schemes to promote afforestation and management strategies, which help the states in developing specific afforestation and eco-development packages for augmenting biomass production through a participatory planning process of Joint Forest Management (JFM).

Community Participation

Keeping in view the objectives of achieving 25 per cent tree/forest cover as part of the monitorable target set out for the Tenth Five-Year Plan, the MEF has launched a very ambitious afforestation programme under the NAEB with people's involvement for the sustainable management of the country's forests. Under this programme, all afforestation schemes of the MEF have been brought under a single National Afforestation Programme being implemented through decentralised Forest Development Agencies (FDAs) set up at the forest division level. The FDAs are a confederation of Joint Forest Management Committees (JFMCs) at the village level to provide an organic link between the forest department and the grassroots level communities.

National Action Programme to Combat Desertification

India is a party to the UN Convention to Combat Desertification (UNCCD) and the MEF is the national coordinating agency for the implementation of the UNCCD in the country. As an affected party, a 20-year comprehensive National Action Programme (NAP) to combat desertification in the country has been prepared.

For the Tenth Five-Year Plan, activities have been initiated that include, among others, assessment and mapping of land degradation, drought monitoring and early warning system groups, drought preparedness contingency plans, and on-farm research activities for development of indigenous technology, etc.

Under UNCCD, a regional action programme for Asian countries has been formulated to strengthen the existing capacity of the member-countries and to network with each other to combat desertification. Six Thematic Programme Networks (TPNs) have been identified for this purpose.

Similarly, India is participating actively in TPN-1 "Desertification Monitoring and Assessment". A pilot project is under progress through the Indian Space Research Organisation (ISRO), Space Application Centre, Ahmedabad, at a cost of Rs 1.39 crore.

India has also joined the TPN-4 "Water Resources Management for Agriculture in Arid, Semi-arid and Dry Sub-humid Areas". The Ministry of Water Resources has been identified as National Task Manger for TPN-4 while TPN-3 on "Range and Pasture Management" has just been launched with Iran as host country. The remaining TPNs, i.e., TPN-5 "Drought Preparedness and Mitigation in the Context of Climate Change" and TPN-6 "Strengthening Planning Capacities for Drought Management and Controlling Desertification" are yet to be launched.

International Cooperation

International Cooperation and Sustainable Development (ICSD) Division is the nodal point within the MEF for United Nations Environment Programme (UNEP), Nairobi, South Asia Cooperative Environment Programme (SACEP), Colombo and the matters relating to sustainable development. The Division also handles bilateral issues and matters pertaining to multilateral bodies such as the Commission on Sustainable Development, Global Environment Facility (GEF) and the regional bodies like Economic and Social Commission for Asia and Pacific (ESCAP), South Asian Association for Regional Cooperation (SAARC), European Union (EU) and the India-Canada Environment Facility.

The MEF is also the nodal agency for various environment-related multilateral conventions and protocols. These include Convention on Wetlands of International Importance, especially as Waterfowl Habitat, Vienna Convention for the Protection of the Ozone Layer, Montreal Protocol on Substances Depleting the Ozone Layer, Conventions on Biological Diversity, UN Framework Convention on Climate Change, Kyoto Protocol, the Basel Convention on Trans-boundary Movement of Hazardous Substances, Convention to Combat Desertification and Stockholm Convention on Persistent Organic Pollutants, etc.

Ozone Cell

Global efforts to protect the ozone layer started in the early seventies leading to the adoption of the Vienna Convention on Ozone Depleting Substances (ODS) in 1985 and the Montreal

Protocol in 1987. India acceded to the Montreal Protocol, along with its London Amendment in 1992. To meet the country's commitment to ODS phase out under the protocol and to disseminate information on ozone and ODS, the MEF has established an Ozone Cell. The MEF provides custom/excise duty exemption for ODS phase out projects and detailed guidelines/procedures have been finalised to grant duty exemptions for new investments with non-ODS technologies. The Reserve Bank of India (RBI) has issued directions to all financial institutions and commercial banks not to finance new establishments with ODS technology. Licensing system has been adopted to regulate import and export of ODS. A ban has also been imposed on trade of ODS with non-parties. The Notified Ozone Depleting Substances (Regulation and Control) Rules, 2000 provide a legal framework for ensuring compliance of the Montreal Protocol. It sets the deadline for phasing out of various ODS, besides regulating production, trade, import and export of ODS. The rules prohibit the use of CFCs in the manufacture of various products beyond 1 January 2003 except in metered dose inhalers and for other medical purposes. Similarly, use of halons is prohibited after 1 January 2001 except for essential uses. Other ODS, such as carbon tetrachloride and methyl chloroform, and CFCs for metered dose inhalers, can be used up to 1 January 2010. Further, the use of methyl bromide has been allowed up to 1 January 2015.

Climate Change

India is party to the United Nations Framework Convention on Climate Change (UNFCCC). The objective of the UNFCCC is to stabilise greenhouse gases concentrations in the atmosphere at a level that would prevent dangerous human induced interference with the climate system. The Convention enjoins upon the parties to communicate to the conference of parties through its Secretariat the following information: (i) a national inventory of anthropogenic emissions by sources and removal by sinks of all greenhouse gases not controlled by Montreal Protocol, to the extent its capacity permit; (ii) general description of steps taken or envisaged by the party to implement the Convention; (iii) any other information that the party considers relevant to the achievement of the objective of the Convention. The MEF is the executing and implementing agency of the project.

India acceded to the Kyoto Protocol in August 2002 with the objective to fulfil prerequisites for implementation of Clean Development Mechanisms (CDM) in accordance with the national sustainable priorities. The Kyoto Protocol commits the developed countries, including economies in transition, to reduce emissions of greenhouse gases by an average of 5.2 per cent below 1990 levels during 2008–12. A National CDM Authority, which has been functioning since 2003, has so far approved 25 projects mainly in the field of renewable energy, municipal solid waste, hydrofluorocarbons (HFCs), small hydro and energy efficiency, etc. It is expected that implementation of these projects would help in attracting foreign investment as well access to more efficient technologies. At the Ninth Conference of Parties (COP-9), held in Milan, Italy in December 2003, the major decisions adopted relate to setting forth the modalities and procedures for the sinks projects in the first commitment period; guidance to the GEF to make operational at the earliest the Special Climate Change Fund; and provide further guidance to the Least Developed Countries Fund. The high-level political debate during the conference took place through three informal ministerial round tables focusing on adaptation, mitigation, sustainable development, technology and assessment. Ministers agreed that climate change remains the most important global challenge to humanity and that its adverse effects are already a reality in many parts of the world. One of the obligations of India under the Convention is to furnish information to UNFCCC Secretariat regarding implementation of Convention decisions.

In India, there is thus a need for sustainable growth, i.e., growth which is not harmful to environment. The eco-friendly development can solve the problem of environmental degradation. Efforts are going on for the environmental management. At all levels of education provisions have been made for the knowledge of environment and its conservation. In the country, many centres are providing special training for environmental management. The programmes of environmental awareness have been launched through media. India is an active member of various international organisations concerning environment. Several programmes are going on under UNEP. The government has recently started emphasising the combined use of regulatory and economic instruments for improving environmental quality. This will be more effective if enforced systematically. The environmental audit introduced in 1993 is a measure to deal with this problem through improved self-regulation. The use of GIS (Geographical Information System) will be useful for effective planning. There is also a need for coordination between government agencies, NGOs and the public for the proper management of environmental quality in the country.

17

Environmental Ethics and Legislations in India

India has inherited a culture of tolerance, non-violence, equity and compassion for animate objects. In the olden times they were a part of daily life and synthesised with the religion. Religious teachings, social and political norms, and economic policies treated man as a part of nature, not as a moulder or superior to it. Air, water, land, animals and human beings are the creation of one supreme power—God. Therefore, the fundamental ethic of behaviour was to live in harmony with each other, because it was well realised that each one of them is dependent on other and destruction or damage to the other is the destruction of self.

Since Vedic times the main motto of social life was "to live in harmony with nature". In spite of the fact that in those days environmental problems were not existing because of lesser population and non-existence of pollution but even then all the great teachers of that time have created an environment of co-existence with nature.

Environmental Ethics in Indian Constitution

The above mentioned philosophy of peaceful co-existence with nature is also enshrined, in one way or another, in the Indian Constitution. Two provisions relating to environment are incorporated in the Indian Constitution in the form of Articles 48A and 51A(g). Article 48A is a constitutional pointer to the state to protect and improve the environment and Article 51A(g) confers a fundamental duty on the citizens of India to protect and improve the environment and have compassion for living creatures. This clearly shows that the Indian Parliament fell in line with old traditional values. The language used in these provisions clearly indicates that the principle of equity, co-existence, reverence for nature and non-violence have been given legal recognition. The use of the term 'protect and improve' implies the improvement of the natural environment as well as quality of life. Further, protection of the environment implicitly directs us not to cut trees, keep the water of rivers, lakes, etc., clean and wholesome. This is reminiscent of centuries old 'Chipko movement' of village Kherjarilli in Rajasthan where Amrita Bai, her four family members and other 359 persons sacrificed their lives to save village trees (popularly known as 'Green Khejris'). Thus, we have a culture where trees are regarded more precious and revered than our lives. The government has also declared the villages of Bishnois in Punjab and Rajasthan as 'reserved areas' and cutting of trees and killing of animals have been declared an offence in these territories. If one happens to visit these villages, he/she can see deers and other innocuous animals roaming there freely and fearlessly.

The phrase "to have compassion for living creatures" used in Article 51A(g) recognises the principle that all creatures are made equal and that animal killing should be prohibited as taught by the principle of non-violence, or as the moral code of conduct says that "killing of animals and birds is a sin of highest order". Abiding by this philosophy of olden times, the Indian Parliament has also passed two statutes, viz., The Wild Life (Protection) Act, 1972 and the Cruelty Against Animals Act, 1962. It is also in consonance with another ethical principle that unarmed enemy and mute birds and animals should not be harmed or attacked. This reminds us of an old story of Prince Siddhartha who saved a swan, who was shot at with an arrow by his brother. The same Prince Siddhartha later became Lord Buddha—a great preacher of non-violence (*ahimsa*) and harmonious relationship with nature. The object behind the theory of "to have compassion for living creatures", scientifically speaking, is that animals and birds are the second trophic level of the 'food chain' or foodwebs and help in energy flow. The utility of the animals and their products need no special description here. It should not also be forgotten that wildlife is the guardian and protector of our forests. Decay of wildlife implies decay of forests. Thus, forests and wildlife are interdependent and complementary to each other.

Indian social and moral ethics, relating to environment protection and conservation, have been accepted by Indian judiciary time and again. Some of the judicial pronouncements upholding old Indian traditions concerning environment are quoted below:

> "Water and rivers have dominated the destiny and fortune of man. Plentiful rivers have brought prosperity to those who lived on their banks. If Bhagirath brought salvation, Ganga sustains life ... ages have rolled up by it and it has remained eternal", (*F.K. Husian vs. Union of India*, AIR 1990 Ker 321)

> "In order to arouse amongst the people the consciousness of cleanliness of environment, the Government of India and the governments of the states and union territories may consider the desirability of organizing 'keep the city clean week', 'keep the town clean week' and 'keep the village clean week' ... at least once a year. During that week, the entire city, town and village should be kept as far as possible clean, tidy and free from pollution of land, water and air". (*M.C. Mehta vs. Union of India*, AIR 1988 SC 1115)

> "Environmental law is based on the realisation of mankind of the dire physical necessity to preserve these invaluable and none too easily replenishable gift of Mother Nature to man and his progeny from the reckless wastage and raparious appropriation that common law permits". (*T. Damondhar Rao vs. Municipal Corporation, Hyderabad*, AIR 1987 SC 171)

The essence of their teachings in the form of *Vedas*, *Upanishads*, *Smritis* and *Dharmashastras* was to protect nature and its elements, i.e., air, water, soil, land, vegetation and animals. In Indian culture, a worshipful attitude has been developed towards trees, earth (*prithvi*), sky (*aakash*), air (*vayu*), water (*jal*), rivers, sun, mountains, etc. It was regarded a sacred duty of every person to protect them. The Hindu religion enshrined a respect for nature, environmental harmony and conservation. Many verses in the *Rigveda* and *Atharaveda* have been devoted to the praise of *Surya Devta* (Sun), *Vayu Devta* (Lord Air), *Agni Devta* (God of Fire), *Varuna Devta* (God of Water), *Prithvi Mata* (Mother Earth), *Vanya Devi* (Goddess of Forests), etc. Therefore, cutting of trees and polluting air, water and land were regarded as sins. Protection of their purity was considered to be the duty of everyone. *Rigveda, Manusmriti* and *Charak Sanhita* have emphasised on purity of water and its healing and medicinal value. The society, as a whole, has developed a code of conduct (*maryada*) to protect nature and live in co-existence with local and regional environment.

Environmental Legislations in India

Before independence, in India, some laws were enacted for the protection of the environment. In Indian Penal Code of 1960, Articles 268, 290, 291, 426, 430, 431 and 432 were related to the environment. Similarly, Article 277 was related to water pollution and Article 278 to air pollution. In Motor Vehicle Act, 1938, there was also a provision to control pollution and Indian Forest Act was passed in 1927. After independence, serious efforts have been made in the form of legislations for the conservation and protection of the environment. Originally, the Indian Constitution did not have explicit reference to environment protection, so there was no independent and separate provision dealing with the protection or improvement of the environment. But, taking note of Stockholm Conference and growing awareness for environmental pollution and eco-imbalances, the Indian Parliament passed a historic amendment—42nd Constitution Amendment Act, 1976. This Act incorporated two significant Articles—Articles 48A and 51A(g)—to protect and improve the environment. Further, it introduced certain changes in the Seventh Schedule of the Constitution. These changes are as follows:

> *Article 48A*: "The State shall endeavour to protect and improve the environment and to safeguard the forests and wildlife of the country."

> *Article 511(g)*: "It shall be a duty of every citizen of India ... to protect and improve the natural environment including forests, lakes and wildlife, and to have compassion for living creatures."

The environmental laws, which have been passed by central and state legislatures from time to time, are based on the recognition of clean environment as a human right or fundamental right. As it has been recognised that clean environment is the basic need for the survival of humanity and it cannot be ensured without ecological balance, thus, this right belongs to all as survival of mankind depends on clean, healthful or pollution-free environment. Any attempt to defile damage the natural environment would amount to violation of human right to clean environment. The Stockholm Conference of 1972 also declared that man has the fundamental right to freedom of equality and adequate conditions of life in an environment of a quality that permits a life of dignity and well-being.

To meet these challenges to mankind, various measures, including legal measures have been adopted in India. Many laws have been passed by the Indian Parliament and the state legislatures to contain the problem of air, water, land, radiation pollution and eco-imbalances. These laws include the Wildlife (Protection) Act, 1972, the Water (Prevention and Control of Pollution) Act, 1974, the Water Prevention and Control of Pollution) Cess Act, 1977, the Forest (Conservation) Act, 1980, the Air (Prevention and Control of Pollution) Act, 1981 and the Environment (Protection) Act, 1986. A brief note on these laws provides a deep insight into the legislative measures taken in India regarding protection of environment.

Environment (Protection) Act, 1986

The Environment (Protection) Act, 1986 was passed by Indian Parliament to fulfil constitutional obligation as provided under Article 48A and also in consonance with Stockholm Declaration of 1972, wherein all governments were expected to evolve necessary laws to protect and improve the flora and fauna, non-renewable resources, wildlife and human health.

Concern over the environment has grown the world over since the sixties. The decline in environmental quality has been evidenced by increasing pollution, loss of vegetal cover and biological diversity, excessive concentrations of harmful chemicals in the ambient atmosphere and in food chains, growing risks of environmental accidents and threats to life support systems. The world community's resolve to protect and enhance the environmental quality found expression in the decisions taken at the United Nations Conference on Human Environment held in Stockholm in June 1972. The Government of India, strongly voiced the environmental concerns. While several measures have been taken for environmental protection in India, both before and after the Conference, the need for a general legislation to further implement the decisions of the Conference has become increasingly evident. Besides, existing laws generally focus on specific types of pollution or on specific categories of hazardous substances. Some major areas of environmental hazards are not covered. There also exist uncovered gaps in areas of major environmental hazards. There are also inadequate linkages in handling matters of industrial and environmental safety. Control mechanisms to guard against slow, insidious build up of hazardous substances, especially new chemicals, in the environment are weak. Because of the multiplicity of regulatory agencies, there is need for an authority which can assume the lead role in studying, planning and implementing long-term requirements of environmental safety and to give direction to, and coordinate a system of speedy and adequate response to emergency situations threatening the environment.

The Environment (Protection) Act, 1986 is a special law that extends to the whole nation. It was passed in March 1986 and came into force on the 19 November 1986. The Act consists of 26 sections divided into four chapters. Seven schedules dealing with emission standards of air, noise, effluents, etc., have also been appended to it. Various rules, including the Environment (Protection) Rules, 1986, have also been provided with it. They are as follows:

1. Hazardous Wastes (Management and Handling) Rules, 1989.
2. Hazardous Micro-Organism Rules, 1989.

3. Manufacture, Storage and Import of Hazardous Chemicals Rules, 1989.
4. Chemical Accidents (Emergency Planning, Preparedness and Response), Rules 1996.
5. Bio-Medical Water (Management and Handling) Rules, 1998.
6. Municipal Solid Wastes (Management and Handling) Rules, 2000.
7. Recycled Plastics Manufacture and Usages Rules, 1999.
8. Noise Pollution (Regulation and Control) Rules, 2000.

According to *Section 2(a)* of the Act, environment includes: (a) water, air and land, and (b) the interrelationship which exists among and between (i) water, (ii) air, (iii) land, (iv) human beings, (v) living creatures, (vi) plants, (vii) micro-organisms, and (viii) property. Thus, it includes both animate and inanimate objects and their interrelationship.

Section 2(b): Environmental pollutant means "any solid, liquid or gaseous substance present in such concentration as may be, or tend to be, injurious to environment".

Section 2(c): 'Environmental pollution' means "the presence in the environment of any environmental pollutant".

Section 2(d): 'Handling', in relation to any substance, means the manufacture, processing, treatment, package, storage, transportation, use, collection, destruction, conversion, offering for sale, transfers or the like of such substance.

Section 3: Power of Central Government to take measures to protect and improve environment.

Section 4: It authorises the Central Government to appoint offices with proper designations.

Section 5: It empowers the Central Government to issue directions to any person/officer/authority appointed under Section 4 of the Act.

Sections 7 to 17: These are related to prevention, control and abatement of environmental pollution.

The Act also empowers the Central Government to constitute a Board of Environment Protection. In fact, this Act is a very comprehensive document on environment protection in India. The other laws on environment have also interconnection with 1986 Act.

Air (Prevention and Control of Pollution) Act, 1981

This Act was passed by the Indian Parliament in the exercise of its powers conferred under Article 253 of the Constitution. The aims and objects of the Act provide:

> "Whereas decisions were taken at the United Nations Conference on Human Environment held at Stockholm in June 1972, in which India participated to take appropriate steps for the preservation of the natural resources of the earth which, among other things, include the preservation of the quality of air and control of air pollution.
>
> And, whereas it is considered necessary to implement the decisions aforesaid in so far as they relate to the preservation of the quality of air and control of air pollution."

The Act aims to achieve the following goals:

(i) to provide for the prevention, control and abatement of air pollution;
(ii) for the establishment of Board with a view to carry out the above mentioned purpose;
(iii) to confer on and assign to such Board powers and functions relating to prevention, control and abatement of air pollution and other matters connected thereto; and
(iv) to lay down the standards to maintain the quality of air.

The term 'air pollution' means the presence in the atmosphere of any pollutant and 'air pollutant' means any solid, liquid or gaseous substance (including noise) present in the atmosphere in such concentration as may be or tend to be injurious to human beings or other living creatures or plants or property or environment. Thus, air pollutants include smoke, smoot, heat, fly-ash, suspended particulate matter (SPM), noise, radioactive substances, etc.

The above definition emphasises two things: (a) high concentration of solid, liquid or gaseous substances, which (b) is injurious or likely to be injurious to human beings, flora and fauna, property or environment. Some small quantity of gases, SPM or vapour is tolerable or cannot be designated as pollutants, unless and until the concentration of such things reaches such a volume that it becomes injurious. For example, sound at low pitch is acceptable and liked, but sound at a very high pitch (known as noise) is not acceptable and becomes a health hazard. The same is true of other gases, liquids and solid substances. A small quantity of pollutants usually does not affect human health adversely. Such quantity or volume may be described as permissible/tolerable limit as the nature also has its self-purification mechanism. But, if the volume or the quantity of the pollutants is such which is deleterious/injurious to the health of human beings, flora, fauna, etc., it becomes environmental pollution. In Taj Trapezium case, the apex court observed that emission of sulphur dioxide from coke/coal-using industries was causing acid rain (the sulphur dioxide when combined with moisture forms sulphuric acid called acid rain) which had a corroding effect on the gleaming white marble of the Taj Mahal. Therefore, 292 industries were ordered either to close down or to switch to using gas.

Sections 3 to 15 of the Act provide for the constitution of a Board, qualifications of its members, terms and conditions of their service, meetings, vacation of seats and temporary association of person with the Board, etc.

The Act envisages two types of Boards—one at the central level and other in the respective states.

Water (Prevention and Control of Pollution) Act, 1974

The main aim of the Act of 1974 is to maintain or restore the wholesomeness of water and to prevent, control and abate water pollution. To achieve these objectives, the Act has provided various provisions, which are very comprehensive. In view of sub-section 2(e), read with Sections 17 and 18 of this Act, the fundamental objective of the statute is to provide clean water to citizens.

Section 2 of the Act defines certain basic terms used in the Act. While defining water pollution, it provides that:

> "Pollution means such contamination of water or such alteration of physical, chemical or biological properties of water or such discharge of any sewage or trade effluent or of any other liquid,

gaseous or solid substance into water (whether directly or indirectly) as may, or is likely to, create a nuisance or render such water harmful or injurious to public health or safety, or to domestic, commercial, industrial, agricultural or other legitimate uses, or to the life and health of animals or plants or of aquatic organism" [Section 2(e)].

Thus, pollution means:

(a) contamination of water, or
(b) alteration of physical, chemical or biological properties of water, or
(c) discharge of sewage or trade effluent, or
(d) any other solid, liquid or gaseous substance, which may or is likely to create
  (i) nuisance, or
  (ii) render such water harmful or injurious to, public health or safety, or to domestic, commercial, industrial, agricultural or other legitimate uses.

Thus, it is a very comprehensive definition and covers all changes in physical, chemical or biological properties of water. The definition also covers the rise in the temperature of water and discharge of radioactive substances in the water.

Further, the Act has used two terms in relation to water pollution: stream and well. The 'stream' here includes (a) river, (b) watercourses (whether flowing or for the time being dry), (c) inland water (whether natural of artificial), (d) sub-terranean water (underground water), and (e) sea or tidal water.

Section 3 provides that a Board shall be appointed and constituted by the Central Government to perform the functions under the Act.

Noise Pollution (Regulation and Control) Rules, 2000

Noise Pollution (Regulation and Control) Rules, 2000 were framed by the Central Government under provisions of the Environment (Protection) Act, 1986 and read with Rule 5 of the Environment (Protection) Rules, 1986. Rule 3 provides for ambient air quality standards in respect of noise for different areas/zones as specified in the schedule annexed to the rules, as under:

Schedule

Ambient Air Quality Standards in Respect of Noise

Area Code	*Category of Area/Zone*	*Limits in dB (A) Leq.*	
		Day-time	*Night-time*
(A)	Industrial Area	75	70
(B)	Commercial Area	65	55
(C)	Residential Area	55	45
(D)	Silence Zone	50	40

Notes: *(1)* Day time shall mean from 6.00 a.m. to 10.00 p.m.

(2) Night time shall mean from 10.00 p.m. to 6.00 a.m.

(3) Silence zone is defined as an area comprising not less than 100 metres around hospitals, educational institutions and courts. The silence zones are zones which are declared as such by the competent authority.

(4) Mixed categories of areas may be declared as one of the four above mentioned categories by the competent authority.

It is for State Government to categorise the area into industrial, commercial, residential or silence zone and to take measures for abatement of noise, including that emanating from vehicular movements. Clause 4 provides that all development authorities, local bodies and other authorities concerned, while planning developmental activity or carrying out functions relating to town and country planning, shall take into consideration all aspects of noise pollution as a parameter of quality of life to avoid noise menace and to achieve the objective of maintaining the ambient air quality standards in respect of noise.

Rule 4: Responsibility as to enforcement of noise pollution control measures:

(1) The noise level in any area/zone shall not exceed the ambient air quality standards in respect of noise as specified in the schedule.

(2) The authority shall be responsible for the enforcement of noise pollution, control measures and the due compliance with the ambient air quality standards in respect of noise.

Rule 5: Restrictions on the use of loudspeaker/public address systems

(1) A loudspeaker or a public address system shall not be used except after obtaining written permission from the authority.

(2) A loudspeaker or a public address system shall not be used at night (between 10.00 p.m. and 6.00 a.m.) except in closed premises for communication within, e.g., auditorium, conference rooms, community halls and banquet halls.

Rule 6: Consequences of any violation in silence zone/area: Whoever, in any place covered under the silence zone/area, commits any of the following offences, shall be liable for penalty under the provisions of the Act:

(i) whoever plays any music or uses any sound amplifiers,
(ii) whoever beats a drum or tom-tom or blows a horn either musical or pressure, or
(iii) whoever exhibits any mimetic, musical or other performances of a nature to attract crowds.

Rule 7: Complaints to be made to the authority:

(1) A person may, if the noise level exceeds the ambient noise standards by 10dB (A) or more given in the corresponding columns against any area/zone, make a complaint to the authority.
(2) The authority shall act on the complaint and take action against the violator in accordance with the provision of these rules and any other law in force.

Rule 8: Power to prohibit, etc., continuance of music, sound or noise:

(1) If the authority is satisfied from the report of an officer in charge of a police station or other information received by him that it is necessary to do so in order to prevent annoyance, disturbance, discomfort or injury or risk of annoyance, disturbance, discomfort or injury to the public or to any person who dwell or occupy property in the vicinity, he may, by a written order, issue such directions as he may consider necessary to any person for preventing, prohibiting, controlling or regulating:
    (a) the incidence or continuance in or upon any premises of
        (i) any vocal or instrumental music,

(ii) sounds caused by playing, beating, clashing, blowing or use in any manner whatsoever of any instrument including loudspeaker, public address systems, appliance or apparatus or contrivance which capable of producing or reproducing sound, or

(b) the carrying on in or upon, any premises of any trade, avocation or operation or process resulting in or attended with noise.

(2) The authority empowered under sub-rule (1) may, either on its own motion, or on the application of any person aggrieved by an order made under sub-rule (1), either rescind, modify or alter any such order.

Wildlife (Protection) Act, 1972

In 1972, the Indian Parliament passed a comprehensive national law—The Wildlife (Protection) Act, 1972—with the sole aim of protecting wildlife. It not only prohibits hunting but also created protected areas and controls trade in wildlife products. To achieve these objectives, it has created a separate and independent authority to protect and improve wildlife. The Act is applicable in all states except Jammu & Kashmir.

The first and foremost purpose of this Act is to protect the habitats of wild animals. As a sequel to it, various national parks and game sanctuaries have been established to ensure greater protection to wildlife. Some special provisions also aim to preserve endangered species like Project Tiger, Gir Lion Sanctuary, Himalayan Musk Deer Project, etc. The Act consists of 60 sections and 6 schedules divided into seven chapters.

Wildlife, according to the 1972 Act, includes any animals, bees, butterflies, crustaceans, fish and moths, and aquatic or land vegetation which forms part of any habitat. This definition is wider in its connotation. The term 'wild animal' has also been defined as any animal found wild in nature and includes any animal specified in Schedule I, II, III, IV or V. Schedule I includes list of mammals (1 to 41-B), amphibians and reptiles (1 to 17-A), birds (1 to 18), crustaceans and insects. It has been clarified that 'animal' includes amphibians, birds, mammals, and reptiles and their young, and in the case of birds and reptiles, their eggs.

Hunting of wild animals has been prohibited altogether. But, the Chief Wildlife Warden has been authorised to grant permission to hunt animals under certain circumstances. These are as follows:

1. if he is satisfied that any wild animal specified in Schedule I has become (i) dangerous to human life; or (ii) is disabled; or (iii) diseased as to be beyond recovery; or
2. when the Chief Wildlife Warden, or the authorised officer, is satisfied that any wild animal, specified in Schedule II, III, or IV, has become dangerous to human life or to property, or is so disabled or diseased so as to be beyond recovery.

The killing in good faith of any wild animal 'in defence' of oneself or of any other person is not an offence.

The Act has also specified provisions to protect specified plants of sanctuaries, national parks, forests or areas specified by notification. All the specified plants, or part or derivation thereof, have been declared to be the property of the state governments, and if they are a part of a sanctuary or national park, they shall be the property of the central government. Therefore, a person must have a licence from the Chief Wildlife Warden, or other authorised officer, to commence or carry on business or occupation as a dealer in specified plants.

The Act also prohibits to (a) wilfully pick, uproot, damage, destroy, acquire or collect, or (b) possess, sell, offer for sale or transfer by way of gift or otherwise, or transport any specified plant, whether dead or alive or part or derivative. But, this provision is not applicable to the member of a Scheduled Tribe.

Similarly, cultivation, dealing in, possession of specified plants without a licence has also been declared to be a punishable act. Further, no person shall purchase, receive or acquire any specified plant, or part or derivative otherwise than from a licensed dealer.

Laws Relating to Forest Conservation

Forest or natural vegetation is an integral element of natural environment. It not only affects but also controls the environment and its degradation. As stated earlier, due importance has been given to forests during ancient and medieval periods but during British rule overexploitation of forests for commercial purposes had begun. But, soon, its importance has been realised and several laws were introduced. In Forest Act, 1865, the forest has been declared as a state government property. The Indian Forest Act, 1927 was enacted with the objective to protect the forests. Under this, forests have been classified into four categories, viz., reserved forests, village forests, protected forests and non-government private forests. This Act was a very elaborate one, which deals not only with the use and conservation of forests but also empowers to punish those who are responsible for the unauthorised cutting of forests.

The Forest (Protection) Act, 1980 was passed in the spirit of the 1972 Stockholm Conference on Environment. This was further modified in 1988. This Act has given elaborate powers to Central Government for the protection and conservation of forests, including biodiversity as well as for the establishment of national parks in consultation with state governments. In order to make this Act more effective, six regional centres at Bangalore, Bhopal, Bhubaneswar, Lucknow, Shillong and Chandigarh have been established. At present, governments (both at Centre and in states) have been very particular about the forest conservation and this culminated in the formulation of National Forest Policy in 1988.

Laws Relating to Hazardous Waste Management

There were various laws, which directly or indirectly dealt with hazardous wastes and toxic substances, but they were not effective under the changing modern conditions. Therefore, the Central Government has passed important rules in exercise of powers conferred by the Environmental (Protection) Act, 1986. These rules are as follows:

1. Hazardous Wastes (Management and Handling) Rules, 1989
2. Manufacture, Storage and Import of Hazardous Chemical Rules, 1989
3. Hazardous Micro-Organisms Rules, 1989
4. National Environment Tribunal Act, 1995

5. National Environment Appellate Authority Act, 1997
6. Bio-Medical Wastes (Management and Handling) Rules, 1989
7. Recycled Plastics Manufacture and Uses Rules, 1999
8. Municipal Solid Wastes (Management and Handling) Rules, 2000

A brief review of some of there rules is presented below:

Hazardous Wastes (Management and Handling) Rules, 1989

These rules aim to deal with the problem of hazardous wastes comprehensively. Although, the term 'hazardous wastes' has not been defined separately in the rules, but Rule 3(i) provides that 'hazardous wastes' means waste substances, which are generated in the processes as indicated in the schedule.

Rule 4 fixes the responsibility of the occupier and operator of the facility for proper collection, reception, treatment, storage and disposal of hazardous wastes without any adverse effect on the environment.

Rule 4A describes the duties of the occupier and operator of a facility as follows:

(a) contain contaminants and prevent accidents and limit their consequences on humans and the environment; and
(b) provide persons working on the site with information, training and equipment necessary to ensure their safety.

'Proper authorisation' by the State Pollution Control Board is really the control mechanism. As per Rule 5, this authorisation letter or permit is issued to a man who is well versed in this technique and possesses proper and adequate facilities, technical capabilities and equipment to handle hazardous wastes safely.

Further Rule 12 provides that there shall be no import and export of hazardous wastes for dumping and disposal. However, as an exception, such wastes may be permitted to be imported only as raw material for recycling or reuse. Such wastes can be used for processing and reuse as raw material only after procuring a proper permit from the State Pollution Control Board. In such a case, the exporting country should also inform and seek permission from the Central Government, which may be granted or refused. The occupier exporting or importing hazardous wastes from or to India shall comply with the Articles of the Basel Convention to which India is a signatory. Further, the Ministry of Environment and Forests shall be the nodal agency to deal with trans-boundary movement of hazardous wastes.

Manufacture, Storage and Import of Hazardous Chemical Rules, 1989

Hazardous chemicals have been dealt with separately under these rules. Following are the requirements which a chemical industry shall be required to fulfil before it starts working or in case of an existing industry within a period of three months of coming into operation of these rules, i.e., November 27, 1989:

1. *Notification of sites*: No one can undertake any industrial activity which would produce a threshold quantity or more of an additional hazardous chemical unless he submits a written detailed report about the industry in accordance with Schedule 7 of the Rules at least three months before commencing the activity (Rule 7). Rules of 1989 have provided a list of 684 such hazardous chemicals. Therefore, an industry before its commencement is required to give the above notice.
2. *Safety report*: Before a chemical industry starts functioning, it has to prepare a safety report containing information specified in Schedule 8 at least three months before commencing that activity and in case of an existing industry within a period of six months after coming into operation of these rules (Rule 10). The report should be prepared with the help of experts not associated with such industrial activity.
3. *Preparation of on-site emergency plan by the occupier*: Under Rule 13, the occupier is required to prepare and keep up-to-date on-site emergency plan furnishing the details of how major accidents will be dealt with. This plan shall include the name of person responsible for safety and who can take action according to the plan. A mock drill based on the on-site emergency plan shall be conducted every six months.
4. *Preparation of off-site emergency plans*: Rule 14 makes it obligatory on the authorities concerned to prepare and keep

up-to-date an adequate off-site emergency plan detailing how emergencies relating to a possible major accident on that site will be dealt with. Further, such plans should be prepared by the authority before a new industry starts its activity or in case of an existing industrial activity within six month of coming into operation of these rules. The authority concerned shall ensure the rehearsal of the off-site emergency plan at least once in a calendar year.

5. *Information to the persons liable to be affected*: One of the important features of the rules is to impose a mandatory duty on the occupier of the industry to take appropriate steps to inform persons outside the site who are likely to be affected by a major accident (Rule 15).

Bio-Medical Wastes (Management and Handling) Rules, 1989

These rules are applicable to all those persons who generate, collect, receive, store, transport, treat, dispose, or handle bio-medical wastes in any form. Bio-medical waste has been defined as "any waste, which is generated during the diagnosis, treatment or immunisation of human beings or animals or in research activities pertaining thereto or in the production or testing of biologicals and including categories mentioned in Schedule I". Schedule I mentions 10 categories of bio-medical wastes, which include human anatomical waste, animal waste, microbiology and biotechnology waste, waste sharps, discarded medicines and cytotoxic drugs, solid waste, chemical waste, incineration ash, etc.

Municipal Solid Wastes (Management and Handling) Rules, 2000

It consists of nine rules and four schedules. Rule 3 (xv) defines 'municipal solid waste' as follows:

> Municipal solid waste includes commercial and residential wastes generated in a municipal or notified area in either solid or semi-solid form including industrial hazardous wastes *but* excluding treated bio-medical wastes.

Rule 3 has also defined various terms used in the rules such as anaerobic digestion, bio-degradable substance, bio-methanation, disposal, land-filling, recycling, leachate segregation, vadose water, vermicomposing, etc.

Rule 4 declares that every municipal authority shall be responsible for the implementation of the rules and for any infrastructural development for collection, storage, segregation, transportation, processing and disposal of municipal solid wastes. The overall responsibility has to be given to the Secretary of the Department of Urban Development of a State/Union Territory and the District Collector/Deputy Commissioner in particular to implement the provisions of the rules.

The Central Pollution Control Board and the State Pollution Control Boards have the responsibility of monitoring compliance with standards regarding groundwater, ambient air quality, leachates and compost quality including incineration standards as provided in Schedules II, III and IV of the rules of 2000. Further, the municipal authority will apply for the setting up of waste processing and disposal facility, including landfills, to the State Pollution Control Board, which shall issue the authorisation within 45 days for a given period. Such authorisation shall be valid for one year only.

The municipal solid wastes shall be managed and handled with the compliance criteria and procedures laid down in Schedule II, which provides parameters of (i) collection of municipal solid wastes, (ii) segregation, (iii) storage, (iv) transportation, (v) processing, and (vi) disposal of municipal solid wastes, and their compliance criteria. Schedule III provides for specifications for landfill sites and Schedule IV lays down the standards for composting, treated leachates and incineration.

National Environment Tribunal Act, 1995

The Act has been passed with a view to provide for 'strict liability' for damages arising out of any accident occurring while handling any hazardous substances, and to dispose of such cases expediously. It has also recognised the principle of 'no-fault liability' in accidents occurring due to hazardous substances. It also provides heads under which compensation for damage may be claimed in such accidents.

Section 3 enunciates the principle of 'no-fault' of the owner to pay compensation for death, injury or damage to human beings and environment. Further, the claimant shall not be required to plead and establish that the death, injury or damage was due to any wrongful act, neglect or default of any person.

The Tribunal will hear the applicant and would be guided by the principles of natural justice and shall also have the power of the civil court as provided under the Civil Procedure Code of 1908 (Section 5). Both the parties shall be heard before an injunction is issued by the Tribunal.

The Central Government shall establish the National Environment Tribunal consisting of one chairperson, vice-chairperson, judicial members and technical members; and the bench shall consist of one judicial member and one technical member each (Section 9). The chairperson either must be or has been a judge of the Supreme Court or a High Court; or has for at least two years held the office of vice-chairperson, who will also be a judicial person (Section 10).

National Environment Appellate Authority Act, 1997

Recently, the Government of India has established a National Environment Appellate Authority which is chaired by a retired Supreme Court judge. It is aimed to hear appeals with respect to restriction of areas in which any industry process/operation of industry shall not be carried out, subject to safeguards provided by the Environment (Protection) Act, 1986. It is a type of check on the establishment of the industries and their likely fall-outs.

Other Acts and Rules

Apart from above mentioned Acts directly related with environment, there are other Acts and Rules also, which are applicable in certain cases. These are:

1. Atomic Energy Act, 1962
2. Radiation (Protection) Rules, 1971
3. Insecticides Act, 1968

4. Indian Fisheries Act, 1897
5. Prevention of Food Adulteration Act, 1954
6. Mines Act, 1952
7. Mines and Minerals (Regulation and Development) Act, 1957
8. Motor Vehicles Act, 1988

In fact, many legislative measures are existing in India for the protection and conservation of environment. There is only need for proper monitoring and implementation of rules and regulations so that our environment becomes healthy and the quality of life of the people of India improves.

Glossary

acid rain A condition in which natural precipitation becomes acidic after reacting chemically with pollutants in the air.

aerosol A substance consisting of small particles, typically having diameters that range from 1/100 micrometer to 1 micrometer.

air pollution The deterioration of the quality of air that results from the addition of impurities.

albedo A measure of the reflectivity of a surface, measured as the ratio of light reflected to light received. A mirror or bright snowy surface has a high albedo, whereas a rough, flat road surface has a low albedo.

arithmetic growth In population studies, growth characterised by the addition of a constant number of individuals during a unit interval of time. For instance, if there are x individuals in year O and x + a in year 1, arithmetic growth implies x + na in year n.

atmosphere The predominantly gaseous envelope that surrounds the earth.

atom The fundamental unit of an element.

atomic nucleus The small positive central portion of the atom that contains its protons and neutrons.

atomic number The number of protons in an atomic nucleus.

autotroph An organism that obtains its energy from the sun, as opposed to a heterotroph, which is an organism that obtains its energy from the tissue of other organisms. Most plants are autotrophs.

background radiation The level of radiation on earth from natural sources.

benthic organism A plant or animal that lives at or near the bottom of a body of water.

biodegradable Refers to substances that can readily be decomposed by living organisms.

biogas Methane produced by fermentation of organic waste. Biogas is chemically identical to methane produced in any other manner.

biomass The total weight of all the living organisms in a given system.

biome A group of ecosystems characterised by similar vegetation and climate that are collectively recognisable as a single large community unit. Examples include the Arctic Tundra, the North American prairie, and the tropical rain forest.

biosphere That part of the earth and its atmosphere that can support life.

birth cohort A group of individuals born in a given period, such as in a particular year.

birth rate The number of individuals born during some period, usually a year, divided by an appropriate population. For example, the crude birth rate in human populations is the number of live children born during a given year divided by the mid-year population of that year.

black lung disease A series of debilitating and often fatal diseases that affect the lungs of miners who work in underground coal mines.

bloom A rapid and often unpredictable growth of a single species in an ecosystem.

branching chain reaction A chain reaction in which each step produces more than one succeeding step.

breeder reactor A nuclear reactor that produces more fissionable material than it consumes.

cancer An abnormal, unorganised, and unregulated growth of cells in an organism.

carnivore An animal that eats the flesh of other animals.

carrying capacity The maximum number of individuals of a given species that can be supported by a particular environment.

chain reaction A reaction that proceeds in a series of steps, each step being made possible by the preceding one. See also branching chain reaction.

chemical bond A linkage that holds atoms together to form molecules.

chemical energy The energy that is absorbed when chemical bonds are broken or released when they are formed.

clear-cutting The practice of cutting all the trees in a designated area. Clear-cutting leaves behind barren, open regions but is sometimes the cheapest way to harvest timber if only short-term economic evaluation is used. See selective cutting.

climate The composite pattern of weather conditions that can be expected in a given region. Climate refers to yearly cycles of temperature, wind, rainfall, and so on and not to daily variations.

climax community A natural system that represents the end, or apex, of an ecological succession.

common mode failure An accident in which a single event causes multiple failures and thereby knocks out redundant systems. See redundancy.

competition An interaction in which two or more organisms try to gain control of a limited resource.

composting The controlled, accelerated biodegradation of moist organic matter to form a humus-like product that can be used as a fertiliser or soil conditioner.

cooling pond A lake or pond used to cool water from an electric generating station or any other industrial facility.

cooling tower A large tower-like structure used to cool water from an electric generating station or any other industrial facility.

critical condition In nuclear science, a critical condition is said to exist when a chain reaction continues at a steady rate, neither accelerating nor slowing down.

critical minimum size In general, a critical condition relates to a point at which some property changes very abruptly in response to a small change in some other property of the system. In ecology, a population is said to be reduced to its critical minimum size when its numbers are so few that it is in acute danger of extinction.

crude rate A vital rate with the entire population of some area as the denominator.

death rate The number of individuals dying during some period, usually a year, divided by an appropriate populations. For example, the crude death rate in human populations is the number of deaths during a given year divided by the mid-year population of that year.

decibel (*dB*) A unit of sound intensity equal to 1/10 of a bel. The decibel scale is a logarithmic scale used in measuring sound intensities relative to the intensity of the faintest audible sound.

demographic transition The pattern of change in vital rates typical of a developing society. The process can be outlined briefly as follows. Birth and death rates in a pre-industrial society are typically very high; consequently, population growth is very slow. Introduction or development of modern medicine causes a decline in death rates and hence a rapid increase in population growth. Finally, birth rates fall, and the population grows slowly once more.

demography The branch of sociology or anthropology that deals with the statistical characteristics of human populations, with reference to total size, density, number of deaths, births, migrations, marriages, prevalence of disease, and so forth.

desert A climate system in which rainfall is less than 25 centimeters per year. These systems are barren and support relatively little plant or animal life.

desertification The diminution or destruction of the biological potential of the land, which can ultimately lead to desert-like conditions.

doubling time The time a population takes to double in size, or the time it would take to double if its annual growth rate were to remain constant.

ecological niche The description of the unique functions and habitats of an organism in an ecosystem.

ecology The study of the interrelationships among plants and animals and the interaction between living organisms and their physical environment.

economic externality That portion of the cost of a product that is not accounted for by the manufacturer but is borne by some other sector of society.

ecology the science of all the relations of all organism and their environment.

ecosystem A group of plants and animals occurring together plus that part of the physical environment with which they interact. An ecosystem is defined to be nearly self-contained, so that the matter that flows into and out of it is small compared with the quantities that are internally recycled in a continuous exchange of the essentials of life.

ecosystem homeostasis The control mechanisms within an ecosystem that act to maintain constancy by opposing externals.

entropy A thermodynamic measure of disorder. It has been observed that the entropy of an undisturbed system always increases during any spontaneous process; that is, the degree of disorder always increases.

environment The sum total of all conditions, agencies and influences which affect the development, growth, life and death of an organism, species or race.

environmental resistance The sum of various pressures, such as prediction, competition, adverse weather, and so on, that collectively inhibit the potential growth of every species.

epidemiology The study of the distribution and determination of health and its disorders.

estuary A partially enclosed shallow body of water with access to the open sea and usually a supply of fresh water from the land. Estuaries are less salty than the open ocean but are affected by tides and, to a lesser extent, by wave action of the sea.

euphotic zone The surface volume of water in the ocean or a deep lake that receives sufficient light to support photosynthesis.

eutrophication The enrichment of a body of water with nutrients, with the consequent deterioration of its quality for human purposes.

expectation of life The number of years an infant can be expected to live under a specified schedule of death rates.

extrapolation The prediction of points on a graphout side the range of observation.

evolution, theory of A theory that states that species are not unchangeable but arise by descent and modification from pre-existing species.

fire climax A condition in which the continuance of a given ecosystem is maintained by fire.

fission (*of atomic nuclei*) The splitting of atomic nuclei into approximately equal fragments.

food chain An idealised pattern of flow of energy in a natural ecosystem. In the classic food chain, plants are eaten only by primary consumers, primary consumers are eaten only by secondary consumers, secondary consumers only by tertiary consumers, and so forth.

food web The actual pattern of food consumption in a natural ecosystem. A given organism may obtain nourishment from many different trophic levels and thus give rise to a complex, interwoven series of energy transfers.

fusion (*of atomic nuclei*) The combination of nuclei of light elements (particularly hydrogen) to form heavier nuclei.

Gaia The ancient Greek goddess of the earth. This word has recently been used to describe the biosphere and to emphasise the interdependence of the earth's ecosystems by likening the entire biosphere to a single living organism.

gas A state of matter that consists of molecules that are moving independently of each other in random patterns.

gasohol A motor fuel consisting of approximately 90 per cent gasoline and 10 per cent ethyl alcohol.

genetic engineering The process of artificially removing specific genes from one organism and replacing them with genetic information from another.

geo-metric growth In population studies, growth such that in each unit of time, the population increases by a constant proportion. Also called exponential growth.

geothermal energy Energy derived from the heat of the earth's interior.

green revolution The realisation of increased crop yields in many areas due to the development of new high-yielding strains of wheat, rice, and other gains in the 1960s.

green-house effect The effect produced by certain gases, such as carbon dioxide or water vapour, that cause a warming of the earth's atmosphere by absorption of infra-red radiation. The term is an inappropriate analogy to green-houses, which were once thought to keep themselves warm by admitting sunlight but retaining infra-red radiation. However, it has been shown that most of the heat retention in green-houses results from the conservation of warm air. The atmosphere, of course, is open, so the mechanism is not the same.

gross national product (*GNP*) The total value of all goods and services produced by the economy in a given year.

half-life (*of a radioactive substance*) The time required for half of a sample of radioactive matter to decompose.

hard energy system An energy system based on large centralised electric-generating facilities. See also soft energy system.

hazardous waste The material that poses a threat to human health and the environment.

heat A form of energy. Every object contains heat energy in an amount that depends on its mass, its temperature, and the specific heat of the materials of which it consists. Heat can also be generated by chemical reactions such as combustion.

heat engine A mechanical device that converts heat to work.

heavy oil Natural petroleum deposits that are too thick to be extracted by conventional pumping techniques. Heavy oils are found in "dry" conventional wells and also in oil shales and tarsands.

hectare A metric measure of surface area. One hectare is equal to 10,000 square metres or 2.47 acres.

herbicide A chemical used to control unwanted plants.

herbivore An animal that eats only plant matter.

heterotroph An organism that obtains its energy by consuming the tissue of other organisms. See also autotroph.

home range The area in which an animal generally travels and gathers its foods.

hormones These compounds can be used as insecticides, for if sprayed properly they will disrupt the life cycle of specific insect larvae and kill them.

humus The complex mixture of decayed organic matter that is an integral part of healthy soil.

hydroelectric power Power derived from the energy of falling water.

hydrological cycle (*water cycle*) The cycling of water, in all its forms, on the earth.

injunction A legal order requiring a defendant to stop doing a wrongful act.

innate capacity for increase The maximum rate of growth of a population as determined by the reproductive capacity of the organism.

inversion A meteorological condition in which the lower layers of air are cooler than those at higher altitudes. This cool air remains relatively stagnant and causes a concentration of air pollutants and unhealthy conditions in congested urban regions.

isotopes Atoms of the same element that have different mass numbers.

joule A fundamental unit of energy, 4.184 joules = 1 calorie.

kinetic energy Energy of motion.

krill Small, shrimp-like crustacean that grow in large numbers in the cold waters of the southern oceans. Krill represent a primary food supply for many species of whales.

legume Any plant of the family Leguminosae, such as peas, beans, or alfalfa. Bacteria living on the roots of legumes change atmospheric nitrogen, N_2, to nitrogen-containing salts that can be readily assimilated by most plants.

mass number The sum of the number of protons and neutrons in an atomic nucleus.

megalopolis An agglomeration of several smaller cities and towns into a single expanded urban area.

meteorology The science of the earth's atmosphere.

mine spoil The earth and rock removed from a mine and discarded because the mineral or fuel content is too low to warrant extraction.

mineral reserves The estimated supply of ore in the ground.

molecule The fundamental particle that characterises a compound. It consists of a group of atoms held together by chemical bonds.

mutagen A substance that causes mutations. An inheritable change in the genetic material of an individual.

mutualism An interaction beneficial and necessary to both interacting species.

natural selection A series of events occurring in natural ecosystems that eliminate some members of a population and spares those individuals endowed with certain characteristics that are favourable for reproduction.

neutron A fundamental particle of the atom, that is, electrically neutral.

nodules Small pieces of rock, typically having a mass of about 1 kg, that are found on the floor of the ocean. These nodules are rich in manganese and also contain several other minerals such as copper, iron, aluminum and cobalt.

noise Unwanted sound.

nuclear reactor A device that utilises nuclear reactions to produce useful energy.

omnivore An organism that eats both plant and animal tissue. Common omnivores include bears, pigs, rats, chickens and people.

open dump A site where solid waste is deposited on a land surface with little or no treatment.

ore A rock mixture that contains enough valuable minerals to be mined profitably with currently available technology.

organic farming A set of farming practices that uses no synthetic fertilisers or chemicals.

organochlorides A class of organic chemicals that contain chlorine bonded within the molecule. Some organochlorides, such as DDT, are effective pesticides. They are generally broad- spectrum and long-lived in the environment.

organophosphates A class of organic compounds that contain phosphorus and oxygen bonded within the molecule. Some organophosphates, used as pesticides, are broad-spectrum and extremely poisonous, although they are not long-lasting in the environment.

Ozone Triatomic oxygen, O_3.

parasitism A special case of predation in which the predator is much smaller than the victim and obtains its nourishment by consuming the tissue or food supply of a larger living organism.

photosynthesis The process by which chlorophyll bearing plants use energy from the sun to convert carbon dioxide and water to sugars.

phytoplankton Any microscopic, or nearly microscopic, free-floating autotrophic plant in a body of water. There are a great many different species that exist in a community of phytoplankton; these plants occur in large numbers and account for most of the primary production in deep bodies of water.

plankton Any small, free-floating organism living in a body of water.

pollution The impairment of the quality of some portion of the environment by the addition of harmful impurities.

polychlorinated biphenyl (PCB) A class of organochloride chemicals that are structurally related to DDT. They were used widely in the plastics and electrical industries until they were found to be potent environmental poisons.

population The breeding group to which an organism belongs in practice. A population is generally very much smaller than an entire species, for all the members of a species are seldom in close proximity to each other.

population distribution The composition of a population categorised by several variables, often age and sex.

population ecology The branch of ecology dealing with the size, growth, and distribution of populations of organisms.

power The amount of energy delivered in a given time interval. Power = energy/time.

prairie An extensive area of fairly level, predominantly treeless land. Prairies are characterised by an abundance of various types of grasses.

predation An interaction in which some individuals eat others.

predator An animal that attacks, kills and eats other animals.

primary consumer An animal that eats plants.

radioactivity The emission of radiation by atomic nuclei.

radioisotope A radioactive isotope.

radiolysis A chemical change initiated by radiation.

rate of natural increase The difference between the crude birth and crude death rates. Also called the crude reproductive rate.

rate of population growth The increase in population over a given time divided by the initial population.

recycling The process whereby waste materials are re-used for the manufacture of new materials and goods.

redundancy Superabundance. In the context of safety systems, redundancy refers to the provision of a series of devices that duplicate each other's functions and that are programmed to go into operation in sequence if a preceding device in the series fails.

rendering The cooking of animal wastes such as fat, bones, feathers, and blood to yield tallow (used in the manufacture of soap) and high-protein animal feed.

replacement level The level of the total fertility rate that, if continued unchanged for at least a generation, would result in an eventual population growth of zero.

resource Any source of wealth or means of producing wealth.

respiration The process by which plants and animals release energy by the step-wise oxidation of organic molecules.

saltwater intrusion The movement of salt water from the ocean to terrestrial groundwater supplies that occurs when the water table in coastal areas is reduced.

sanitary landfill A site where solid waste is deposited on a land surface, compacted, and covered with dirt to reduce odours and prevent disease and fire.

savannah A type of tropical or sub-tropical prairie that is subject to seasonal patterns of rainfall. Savannahs are common in central Africa.

secondary consumer A predator that eats an animal that eats plants.

selective cutting The process of cutting selected trees in a stand of timber. See clear-cutting.

smog Smoky fog. The word is used loosely to describe visible air pollution.

soft energy system An energy system based on the use of small, home-sized, decentralised devices using renewable resources.

solar cell A semi-conductor device that converts sunlight directly into electrical energy.

solar collector A device designed to concentrate solar energy for a useful purpose.

solar design active systems: Heating systems that use a controlled flow of some substance; usually air or water, to collect, store, and transmit heat from the sun. *passive systems*: Construction systems that use the structure of a building itself to collect, store, and transmit heat from the sun.

solar energy Energy derived from the sun.

species A group of organisms that interbreed with other members of the group, but not with individuals outside the group.

strip mining Any mining operation that operates by removing the surface layers of soil and rock, thereby exposing the deposits of ore to be removed.

subsidence The settling of the surface of the ground as deep ground water or minerals are removed.

subsidies Economic incentives to induce polluters not to despoil the environment.

succession The sequence of changes through which an ecosystem passes during the course of time. *Primary succession* is a sequence that occurs when the terrain is initially lifeless, or almost so. *Secondary succession* is the series of community changes that takes place in disturbed areas previously occupied by a biological community synfuels An abbreviation for synthetic fuels, liquid fuels produced from oil shales, tar sands, or coal.

sustainable development The development that meets the needs of the present without compromising the ability of future generations to meet to meet their own needs.

synthetic chemicals Compounds produced by chemical reactions performed in a laboratory, as opposed to those formed naturally.

technological fix An approach to solving a social problem, such as environmental degradation, by technical or engineering methods.

thermal pollution A change in the quality of an environment (usually an aquatic environment) caused by raising its temperature.

thermodynamics The science concerned with heat and work and the relationships between them. *First Law of Thermodynamics*: Energy cannot be created or destroyed. *Second Law of Thermodynamics*: It is impossible to derive mechanical work from any portion of matter by cooling it below the temperature of the coldest surrounding object.

threshold level The minimum dose of a toxic substance that causes harmful effects.

tidal energy Energy derived from the movement of the tides.

total fertility rate (*TFR*) The total number of infants a woman can be expected to bear during the course of her life if birth rates remain constant for at least one generation.

toxic substance Any substance whose physiological action is harmful to health.

trespass An illegal intrusion or invasion of other people's property.

tropical rain forest A biome that occurs when high, fairly constant rainfall and temperature permit plants to grew rapidly throughout the year.

tumour An abnormal growth of unregulated cells.

tundra Arctic or mountainous areas that are too cold to support trees and are characterised by low mosses and grasses.

urban sprawl The growth of suburban areas and the expansion of cities to pre-empt increasingly large areas of land.

urbanisation A demographic process characterised by movement of people from rural to urban settlements, from small towns to large cities, and from large cities to their suburbs.

vital event In demography, a birth, a death, a marriage, a termination of marriage, or a migration.

water pollution The deterioration of the quality of water that results from the addition of impurities.

water table The upper level of water in the zone of saturated subsurface soil and rock.

zero population growth A condition in which the birth and death rates in a population are equal therefore, the size of the population remains constant.

Bibliography

Arvill, R. (1967) *Man and Environment: Crisis and the Strategy of Choice*, Penguin.

Allen, J.P. (1991) *Biosphere 2: The Human Experiment*, Penguin, New York.

Bach and Wilfried (1972) *Atmospheric Pollution*, McGraw Hill, New York.

Bahuguna, S. (1990) "Tehri-The Dam of Discontent", *Yojana*, Vol. 34 (10) 9–12.

Barrow, C.J. (1995) *Developing the Environment: Problems and Management*, Longman, Harlow.

Brown, E.H. (1970) "Man Shapes the Earth", *Geographical Journal*, Vol. 136, 74–85.

Brij Gopal and N. Bhardwaj (1979) *Elements of Ecology*, Vikas Publications, New Delhi.

Burton, Robert, W.K. and Gilbert, F.W.(1978) *The Environment as Hazard*, Oxford University Press, New York.

Billihgs, W.D. (1971) *Plant, Man and the Ecosystem*, Macmillan.

Canter, L.W.C. (1996) *Environmental Impact Assessment*, McGraw-Hill, New York.

Clapham, W.B. (1973) *Natural Ecosystems*, Macmillan, London.

Colinvaux, Paul A. (1973) *Introduction to Ecology*, John Wiley and Sons, New York.

Darlingten, P.J. (1957) *Zoogeography*, John Wiley, New York.

Dasseman, R.F. (1976) *Environmental Conservation*, John Wiley, New York.

Detwyler, T.R. (1971) *Man's Impact on Environment*, McGraw-Hill, New York.

Dogra, Bharat (1980) *Forest and People*, Himalaya Darshan Prakashan Samiti, Rishikesh.

Erocal, D. (ed.) (1991) Environmental Management in Developing Countries, Development Centre, OECD, Paris.

Farmer, A. (1997) *Managing Environmental Pollution*, Routledge, London.

Flanagan, D. (1970) *The Biosphere*, W.H. Freeman and Co., San Francisco.

Furley, P.A. and Newey, W.W. (1983) *Man and the Biosphere*, Butterworths, London.

Futehally, I. (1988) *Our Environment*, N.B.T., New Delhi.

Gerasimov, I.P. (1983) *Geography and Ecology*, Progress Publishers, Moscow.

Glan, L.A. and Rhoda, L.M. (eds.) (1970) *Ecological Crisis Readings for Survival*, Harcourt Brace Jovanovich, New York.

Haggett, P. (1972) *Geography: A Modern Synthesis*, Harper and Row, London.

Hunt, D. and Johnson, C. (1995) *Environmental Management Systems*, McGraw Hill, London.

Irving, R.M. and Priddle, G.B. (1971) *Crisis-Reading in Environmental Issues and Strategies*, Macmillan, London.

Jackas, G.V. and White, R.O.(1939) *The Rape of the Earth*, Fabev, London.

Kneese, A.V. (1977) *Economics and the Environment*, Penguin.

Lowe, J. and Lewis, D. (1980) *The Economics of Environmental Management*, Oxford.

Maddox, J. (1972) *The Doomsday Syndrome*, McGraw-Hill, New York.

Manivasakam, N. (1992) *Environmental Pollution*, N.B.T., New Delhi.

Mash, G.P. (1964) Man and Nature (*Physical Geography as Modified by Human Action*) Charles Scriber, New York.

Maxwell, K.E. (1973) *Environment of Life*, Dickenson Publishers, Co., California.

McCormick, J.F. (1992) *The Global Environment Movement*, Belhaven, London.

McGill, S. (1999) *Environmental Risk Management*, Routledge, London.

Nicholson and Max (1970) *The Environmental Revolution*, McGraw-Hill, New York.

Nijkamp, P. (1980) *Environmental Policy Analysis*, Wiley, New York.

Odum, Eugene P. (1971) *Fundamentals of Ecology*, W.B. Saunders Co., London.

Osborn, F. and Bigg, T. (1998) *Earth Summit II: Outcomes and Analysis*, Earthscan, London.

Owen, L. and Unwin, T. (eds.) (1997) *Environmental Management*, Blackwell Pub., Oxford.

Park, C.C. (1980) *Ecology and Environmental Management*, Butterworths, London.

Pfajjlin, J.R. and Ziegler, E.N. (eds.) (1976) *Encyclopedia of Environmental Science and Engineering*, Vols. I and II Gordon and Breach Science Publishers, New York.

Rao, K.L. (1975) *India's Water Wealth*, Orient Longman, New Delhi.

Redclif, M.E. (1987) *Sustainable Development*, Methuen, London.

Rodda, A. (1991) *Women and Environment*, Zed Press, London.

Saxena, H.M. (2004) *Environmental Geography,* Rawat Pub., Jaipur.

Sandgrove, K. (1992) *The Green Manager's Handbook*, Gower, London.

Samson, F.B. and Knopf, F.L. (eds.) (1996) *Ecosystem Management: Selected Readings*, Springer, New York.

Saraf, R.K. (eds.) (1973) *Proceedings of Symposium on Environmental Pollution*, CEPHERI, Nagpur.

Seshadri, B. (1969) *The Twilight of India's Wild Life*, Oxford, London.

Sewell, G.H. (1975) *Environmental Quality Management*, Prentice-Hall.

Singh, H.H. and Others (eds.) (1986) *Geography and Environment*, Concept, New Delhi.

Stephen Tinsley (1999) *Environmental Management Plans Demystijied,* Spon Press, London.

Sharma, H.S. (eds.) (1988) *Environmental Design and Development*, Scientific Publications, Jodhpur.

Sharma, P.D. (1988) *Elements of Ecology*, Rastogi, Meerut.

Singh, L.R. and others (eds.) (1983) *Environmental Management*, Allahabad Geographical Society, Allahabad.

Singh, Savindra and Tiwari, R.C. (eds.) (1989) *Geomorphology and Environment*, Allahabad Geographical Society, Allahabad.

Singh, S.P. (1990) "Tehri Dam: An Assessment and Vision", *Yojana*, Vol. 34 (10) 5–8.

Smith, G.H. (ed.) (1965) *Conservation of Natural Resources*, New York.

Stoddard, D.R. (1965) "Geography and the Ecological Approach", *Geography*, Vol. 50, 242–51.

Strahler, A.N. and Strahler, A.H. (1976) *Geography and Man's Environment*, John Wiley, New York.

Thomas, W.L. (1965) *Man's Role in Changing the Face of the Earth*, University of Chicago Press.

Turk, J. (1985) *Introduction to Environmental Studies*, C.B.S. College Publishers, Chicago.

UNDP (1992) *Handbook and Guidelines for Environmental Management and Sustainable Development*, New York.

UNESCO (1981) *M.B.A. Information System Biosphere Reserves*, Publishers, No. 2.

UNESCO (1970) *Use and Conservation of Biosphere*, Natural Resources Research 10, Paris.

UNO (1992) *Agenda 21: Programme of Action for Sustainable Development,* UN Publication.

Watt, K.E.F. (1969) *Ecology and Resource Management*, McGraw Hill, New York.

Whitehead, M. (2007) *Spaces of Sustainability,* Routledge, London.

Zimmermann, E.W. (1951) *World Resources and Industries*, Harper and Row, London.

Index